Beyond Spring

Wanderings through Nature

MATTHEW OATES

Fair Acre Press

First published in Great Britain in 2017 by Fair Acre Press
www.fairacrepress.co.uk

BEYOND SPRING *Wanderings through Nature* © Matthew Oates 2017

Typeset in Optima and Gill Sans by Nadia Kingsley
Internal design by Nadia Kingsley

Printed and bound by Lightning Source.

A CIP catalogue record for this title is available from the British Library.

ISBN 978-1-911048-23-7

Front Cover Image by Nadia Kingsley
Back Cover Image by Matthew Oates
Cover design by Nadia Kingsley

To the memory of the long hot summer of 1976

Acknowledgements

The publishers gratefully acknowledge permission to reprint copyright material
in this book, in part or in its entirety, as follows:

'Coleshill' in part, from *Coleshill* by Fiona Sampson
(Chatto & Windus, 2013).

'The day after' in part, from *Road Kill* by Nadia Kingsley
(Fair Acre Press, 2012).

'Brimstone' in part, from *Shropshire Butterflies - A poetic and artistic
guide to the butterflies of Shropshire* by Keith Chandler
(Fair Acre Press, 2011).

'Rooks over Avebury' in its entirety, from *Kith* by Jo Bell
(Nine Arches Press, 2015).

'A Spring Wife' in part, from *The Invisible Gift* by David Morley
(Carcanet, 2015).

'Buzzard Soaring' in part, from *Selected Poems* by Roger Garfitt
(Carcanet, 2000).

'Bluebells' in part, from *The Remedies* by Katharine Towers
(Picador, 2016).

'Skylark' in its entirety, from *The Lost Music* by Katrina Porteous
(Bloodaxe Books, 1996).

'Chorus' in part, from *The Invisible Gift* by David Morley
(Carcanet, 2015).

'Now, Under the Trees' in part, from *Hide* by Angela France
(Nine Arches Press, 2013).

Extract from *The Three Winds by* Laurie Lee (Laurie Lee Estate).

'Hay' in part, from *Cure for a Crooked Smile* by Chris Kinsey
(Ragged Raven Poetry, 2009).

Extract from *Coot Club* by Arthur Ransome (Published by Jonathan
Cape. Reprinted by permission of The Random House Group
Limited. © Arthur Ransome, 1934).

'Backwaters: Norfolk Fields' in part, from *An English Apocalypse* by
George Szirtes (Bloodaxe Books, 2001).

'Purple Hairstreak' in part, from *Shropshire Butterflies - A poetic and
artistic guide to the butterflies of Shropshire* by Jean Atkin
(Fair Acre Press, 2011).

'The Ascent of Kinder Scout' in part, from *Due North* by Peter Riley
(Shearsman Books, Bristol, 2015).

'Summer 1976' in part, from *A Horse Called House* by Jonathan
Davidson (smith|doorstop, 1997).

'Swallows' in part, from *Urban Birds blog*, by Keith Chandler
(Fair Acre Press, 2017).

'August' in its entirety, from *The Remedies* by Katharine Towers
(Picador, 2016).

'Steep' in its entirety, from *Edward Thomas Fellowship Newsletter* by
Matthew Oates (2015).

The publishers extend sincere thanks to the following poets for their kind and generous permission to print their previously unpublished material, in part or in its entirety, as follows:

Alison Brackenbury for 'The country writers (Edward Thomas, Richard Jefferies and others)', and 'Why' in part; for 'Spring comes when', 'Territories', and 'Postcard' in their entirety.

Chris Kinsey for 'A Charm of Goldfinches' in its entirety.

Nadia Kingsley for 'Summer Rain' in part.

Matthew Oates for 'Beyond Spring' in its entirety.

***

This book's roots reach back to my old school, Christ's Hospital in West Sussex, and the English tutors there. I am forever grateful to the school and its tutors, most notably to Peter J Cornish who stimulated my lifelong interest in poetry and, somehow, helped a wayward soul develop the confidence to write. He also introduced me to Edward Thomas, whilst my late mother Helen Martin Oates introduced me to WH Hudson's writing. My dearest friend from schooldays Dr Nigel Fleming has encouraged me throughout. I'm not sure why.

I am deeply grateful to Jonathan Bate for his inspirational work on poetic-nature *The Song of the Earth*, a book which changed my life by making me realise the importance of the poetic approach; to Richard Holmes for his seminal works on my lifelong hero Samuel Coleridge; and The Friends of Coleridge, The Edward Thomas Fellowship, the Romanticism Blog run by The Wordsworth Trust at Grasmere, and my close friends within the Brokenborough Poets group; and also to Heather Cobby for constructive comments on my chapter about the Dymock Poets, and Peter and Rita Boogart of the

Netherlands, for their book *A272 An Ode to a Road*, which has run to four editions and champions a road which has enchanted me since I first travelled it, in September 1963.

My sincere thanks to Nadia Kingsley for her editing support, and to Annie Wilson for her careful proofreading and copy-editing of this book. Also to Dr Helen Roy of the Centre for Ecology & Hydrology for helping with my account of the 1976 ladybird invasion.

The National Trust cares for several of the places and landscapes featured in this book: Arnside Knott in Cumbria, Avebury in Wiltshire, Borrowdale in Cumbria, and much of the Lake District high fells, Cherhill Downs & Calstone Coombes in Wiltshire, Countisbury in North Devon, the Manifold Valley & Dovedale in the Peak District, the northern Quantocks in Somerset, Silverdale in Lancashire, Watlington Hill in Oxfordshire, The White Cliffs of Dover in Kent, and Woolacombe in North Devon. The Trust's Spirit of Place programme helped me enormously, by making me begin to realise the importance of our relationship with place.

Equally, I owe a huge debt of gratitude to Charlie and Issy Burrell at the Knepp Castle Estate in West Sussex, and to their staff there. Many passages of this book were written there. It's that sort of place.

Last but by no means least, my dear wife Sally, who has once more tolerated my mental absence for long periods. My children have also suffered. Writing a book must surely be the most selfish thing one can do.

Matthew Oates, May 2017

Contents

Beyond Spring

It had rained more recently up here
Than down along the sunken old way,
And birds sang with greater resonance;
They revelled in newborn air, becoming
Part of it, so that it sang through them.

Backlit by diamond shafts of April light,
Cascades of pearling droplets sudden fell
From families of soft changeling leaves;
Each one in yellow-greens unnamed,
Greening within an ecstasy of opening.

And we, blessed and unblessed with souls,
Cannot love those spellbound moments
Deep enough: they feel far too profound
For our modern manikin world, and hang
Around our necks, guilt-gilded, as memories.

Matthew Oates
Hailey Wood, Gloucestershire, 23rd April.

Before the Beginning

Much have I travell'd in the realms of gold

John Keats - *On First Looking Into Chapman's Homer*

Beyond Spring was inspired by, and perhaps almost written by, some of the wonderful places I found myself in during the brighter months of an English year. All I did was wander about, or loiter with intent, in special places in the great outdoors, and observe what was going on around me, and noted down what was happening. It may even be that some of those places, and perhaps nature itself, used me as a conduit.

Beyond Winter might have been a more appropriate title, but sounded a trifle negative. Whatever, this book is a love affair, an entirely natural love affair – with nature, and with some of the places and times which I've been privileged to experience.

The book's essential subject matter is the experience of spring and summer in the natural world, in England, from the differing perspectives of poetics, traditional natural history, and the evidence-based approach of modern ecology. Our experiences of course, take place within the contexts of time and place, so this book is concerned with love of time and place too, through the stained-glass windows of two of our seasons. It describes fragments of a journey through time, from the genesis of spring to summer's ending – as undertaken by someone who is simply a pilgrim in the world of nature.

The journey meanders over much of England. Essentially, though, it is not offered up as *my* journey but the reader's; so much so that most chapters avoid or minimise usage of the first person singular.

Please do not read this book at the wrong time of year – get out there and live spring and summer instead, and save the book for the darker months, for the book is written in part to help people survive the morbidity of winter. But it is written primarily to help the many lovers of nature who are entrapped in the world of materialism to escape back into the real world, the one in which we belong – the deeply magical world of nature. Perhaps it should have been entitled *Up Yours, Winter!* or even *Beyond Human Resources*. Please read it during lunch breaks in soulless offices, sneakily beneath a pile of papers during pointless office meetings, and especially on dire and repetitive commuter train journeys. Please read it brazenly on the London Underground: it may help you survive there. Most of the chapters are short enough for snatch-reading.

These chapters have been inspired by the great Victorian and Edwardian nature writers, Richard Jefferies, WH Hudson and Edward Thomas; by one chapter of rampant pantheism in Kenneth Grahame's *The Wind in the Willows*; by the Romantic poets' metaphysical relationship with nature; and equally by today's conservation scientists with their ever-evolving ecological knowledge; and the phenomenal outpouring which has been labelled 'new nature writing'. Perhaps art, science and spirituality are at their strongest when they combine, which they do around nature.

One of the main themes is memory, and in particular how our

experiences of nature pass into memory, and transmogrify. But memory is not something confined to the human mind – for places seem to collect people's memories, somehow, and retain them as their own. I do not understand this, which can loosely be termed the collective memory of place, but am convinced that it is real. It is explored here. Places use us, and situations. Poets at times understand this, which helps make them relevant.

Although the bulk of the adventures and experiences described here occurred over a seven-month period during the sunnier side of one particular year they are not necessarily rooted there (some pieces were penned in preceding years). They are simply *of spring* or *of summer*, or both. Therefore, this book is a tribute to Spring and Summer more generally.

It is also a tribute to the poet, naturalist and landscape writer Edward Thomas, a hundred years on from his death on the Western Front. What matters about Thomas is not so much what he did or said, or even wrote, but his way of looking at and dwelling within the world. That particular torch needs carrying forward.

As a benchmark or touchstone, some forty years on, there are reflections on, and even excursions into, that most glorious of summers of modern time – the one that became known as 'the long hot summer of 1976', that *annus mirabilis* – to which this book is dedicated.

Composed in, or even by, Overley Wood, near Cirencester, Gloucestershire, 8th April.

They scared a hundred sparrows from eared corn,
woke to their thrushes' common song.
The line of rooks which flared the sky to roost
stretched five miles long.

First, all they wanted was to tramp all day
out in the air, though soaked or cold.
They rented rooms; but the old farmer died.
Those hills were sold.

Were woods, the wren in bedroom, true, or just
a dream behind a lost child's eyes?
Today I saw a hundred peacocks whirl,
swallows fill skies,

I did not dream of butterflies, but sought
in hot town nights, before first engines' hum,
my visa for your country to which I
have not yet come.

Alison Brackenbury - *The country writers (Edward Thomas,
Richard Jefferies and others)*

At Candlemas

It began with a breath, sent from afar. The breath became a gentling wind, to start the gradual process of drying out the land and relieve the world from winter's soggy grip. It set the lark to ascend, for that first time, from the ploughlands below the heights of Cherhill Downs, just outside but several dimensions distant from the bustling town of Calne on the North Wiltshire Downs, where the poet-metaphysician Samuel Coleridge once dwelt.

There it rattled the latch of a loose bridle gate, which often swings and sings up there in the wind under differing skies but always in the wind, up there on Cherhill's wind-kept summit. There, along the downland crest, beneath a hedgehog clump of stunted trees whose size belittles their age, the wind dried out the topmost layer of last year's fallen Beech leaves, huddled in sunken dells and telling of a time bygone and spent, and issued the message that now is the time for renewal.

And so the song-dream was born from a breath that became a southing breeze before veering a day or so later into a desiccating easterly, to start the journey into spring. The Meadow Pipits knew it for what it was, a calling; or so it seemed to a lone man lurching his way up the sunken chalk-mud track that slowly ascends Cherhill. The place itself knew, perhaps, that the Wheatears homeward journey would soon commence on a warming southerly; for the Wheatear loves this lonely wind-swept down and its secret chalkland sister combes, the dream-land of Calstone, to the south.

The Wood Pigeons, so long misanthropic, recognised the call-ing too, and set to feast on Ivy berries in a roadside hedge, wings pressed clumsily downwards into the foliage, like a Sparrowhawk hunched over its prey. They were struggling to balance on foliage too weak to support their weight but were there, for time had sweetened the berries into winter's final fare, and the song-dream had come alive, even along the old coaching road that has become the roaring fast-track petrol-headed A4.

The white claggy chalk marl, where footfalls suck or slip at random, led the man to the Down's summit where he could lean against a loose-latched gate and dream a while of spring – for the song-dream had reached him too – now that he had escaped the traffic noise. He saw that seagulls had drifted inland, seeking to follow the plough, for spring ploughing was beginning. It seemed that the heads of some of the Black-headed Gulls were starting to blacken, patchily, as they do before the breeding season calls them. The gulls formed a spiral vortex above a solitary green and yellow tractor, out ploughing a waste of stubble land, and by doing so they called in other gulls. The Rooks, ever keen-minded, saw it too and called more of their own, for the worms were rising.

But the song-dream did not just begin here, on Cherhill's high Down, but at a myriad other places deep within the south country, in diverse yet harmonised ways. This genesis was not a feature of any particular place, but of time; though one needed to be in a place that seemed right, and in a rightful mind, to perceive and receive that precious moment. Even though the sun soon became lost to grey-headed clouds, ragging in from nowhere, the banishment of winter had begun.

This was Candlemas, which traditionally occurs on February 2nd and supposedly marks the half-way point of our winter. That may once have been the case, but in modern times February is usually the last month of winter, or even the transitional month from winter to spring, so Candlemas tends nowadays to mark the two-thirds or three-quarters stage of winter. It is a Christian festival associated with light. It is big in some other countries, and within the Roman Catholic Church. Of course it may have hijacked an older pagan festival here, though there is little evidence for that. Whatever, it is a word of profound beauty with poetic and atavistic connotations. It instils a sense of longing, for spring, and for the ending of winter's tyranny. It offers hope, through the lighting of the lamp, and the genesis of the song-dream of summer long.

With impeccable timing, a Fieldfare flock settled amongst ash-grey boughs, heads pointing windward, then silently moved off north, knowingly.

Cherhill Downs, North Wiltshire, early February.

We're not built

but become: trembling columns
of apprehension that ripple
and pass those ripples to and fro
with the world that shakes around us —
it too is something poured
and ceaselessly pouring itself.

February shakes the fields
and trembles in each yellow willow.

Fiona Sampson - *Coleshill*

A Eulogy to the Rook

Over the land freckled with snow half-thawed
The speculating rooks at their nests cawed
And saw from elm-tops, delicate as flower of grass,
What we below could not see, Winter pass.

Edward Thomas - *Thaw*

In England, nature's New Year's Day must be the day when the Rooks start to build or rebuild their nests in earnest. Sure, there are early, mainstream and late-starting districts, each with early and late rookeries, but most lowland and southern rookeries fit into the mainstream category and kick off mostly around Valentine's Day. The birds wait until the temperature starts to warm up, the winter rains ease and, crucially, for when the land starts to dry out. Then they begin, early birds first. This is a highly salient event to modern society, for many of today's rookeries are situated along roads – one of the mightiest in the country is along the M4, mid-way between Reading and Newbury, off the eastbound carriageway. Motorway service stations are a delight to them, offering fast food easy pickings.

Rooks in early spring are one of our top wildlife wonders, and not just around the rookeries, for their antics in the open fields are not so much comical – for that is completely the wrong way of looking at Rooks – as genuinely fascinating, and wondrous to behold. Rooks are, though, at their very best when caught within the whirling dervish of a mad March sky. Then, these ragged but agile birds gyrate a dance integral to the coming of spring, accompanied by cries as varied as the early spring sky itself – no two utterances of *Caw!* are alike, it is a diverse language. Rooks are the high spirits of that time, the time that sees the change from winter to spring.

Perhaps we don't marvel at Rooks enough because we have become too obsessed with rare and declining species, and so undervalue the more commonplace – until that too starts to vanish, and we are tipped off by the doom and gloom sector of the nature conservation movement. The goal then is quite simple: we need to develop *Rookery Tourism* as part of the rural diversification programme, and raise the profile of this underrated and rather vilified bird within our culture. We need to go out rooking and establish the Rook as an accepted cultural icon, ranked alongside the Nightingale and Skylark, Bluebell woods, the leaping Salmon and the rutting Red Deer stag. Rooks reach the parts other birds don't reach, not so much through their intriguing behaviour but through the way in which they epitomise the intense magic of the transition from late winter to early spring.

It was an early March trip to Salisbury Plain that finally made me see the light. Never mind the scientific term 'metapopulation' (an interconnected cluster of colonies ebbing and flowing over areas of landscape) for the Plain is a veritable Rook empire. Rookeries are frequent, almost commonplace, along the five river valleys that divide up and surround the Plain, especially around villages. Watching and listening to nest building in such places is a major experience, intellectually and spiritually – though some people may find living next to a rookery a trifle difficult in March as the cawing starts before first light. Consequently, I ignored the Goosander that flew past me over the River Wylye near Steeple Langford, and several Little Egrets. My task was to venture deeper.

Moving on to the Plain itself, under a lead-white sky, I found that Rooks abound over the vast cornfield wastes, or plough-lands as they used to be called. These arable expanses are nothing new: in WH Hudson's eulogy to old Salisbury Plain *A Shepherd's Life* (1910), he writes of a 'desolate scene', for 'the land was all ploughed and stretched away before me, an end-less succession of vast grey fields, divided by wire fences'. Since Hudson's day the fields have expanded, and the farmland songbirds and arable weeds he knew have all but vanished. But ask any Rook and they will tell you that this is a feeding paradise. Even better for them are the huge open-air pig farms that are scattered about on the Plain. These move about from place to place, in eras, and with them move some rookeries. Between Chitterne and Tilshead I ignored an exultation of ethereal larks and my first Corn Bunting in an embarrassingly long while – both outgunned by the antics of myriad Rooks, many of which were busily perched on the backs of wallowing sows. I had entered the time and place of the Rook.

Above all, Rooks are an integral part of the immense sense of *genius loci* that pervades throughout the 300 square miles of Salisbury Plain. The Plain is a brooding place, with its own climate and towering, often threatening skyscapes, for clouds gather ominously over it in all seasons. It is a place of frequent, sudden, and often extremely loud reverberations for it is used intensively for tank and artillery training. In its centre is some-thing the MOD calls the Central Impact Zone, where live shells explode. At times the Plain threatens to produce mirages, like the great plains of Hungary and Eastern Europe. In cold winters it becomes snowbound and cuts itself off from the rest of the country. In hot summer weather it generates its own thunderstorms. WH Hudson, again in *A Shepherd's Life*, records an intense feeling of 'emptiness and desolation, which frightens the stranger'. Edward Thomas, in *In Pursuit of Spring* (1914) goes further, describing the Plain as a 'sublime inhospitable wilderness' that is haunted by Rooks, and stating: 'The most numerous things on the Plain are sheep, rooks, pewits and larks' – only the sheep and the Pewits are now gone. Thomas goes deeper still: 'It makes us feel the age of the earth, the greatness of Time, Space and Nature.' Indeed, it reeks of history. It is history. But the Plain offers the modern naturalist a feeling of overwhelming loss, for it is our sole surviving tract of rolling downland on deeper soils. Elsewhere on the chalk-steep slopes, where the soil is thin and worthless, the odd isolated fragment survived the ravages of the twentieth-century agricultural revolution. It was military use that saved Salisbury Plain from this fate, together with its smaller cousin to the south-east, Porton Down.

One positive change that has occurred within the countryside is that Rooks are no longer mercilessly shot. As a child in west Somerset I was horrified, and petrified, by the frequent sight of Rook carcasses hanging bedraggled on barbed wire fences, like machine-gunned soldiers after the first day of the Battle of the Somme, and also by rotting rags of black feathers dangling, like hanged men, from farm string attached to poles angled into the ground. This may have as much to do with the banning of cheap homemade cartridges a while back, as with the grudging recognition that Rooks do more good than harm, devouring pests like Eelworms. But it is nonetheless welcome. Also, it appears that people have at last seen through the rural myth that Rooks will desert a rookery unless they are periodically shot – a belief that makes no ecological sense whatsoever, and which was perpetuated solely to justify carnage, for many people's relationship with nature stares over the barrel of a gun, incongruously.

But Rooks are not without sin. They have some despicable habits, and not merely the practice of defecating on vehicles foolishly parked beneath their rookeries. In March, they have the infuriating habit of invading farm granaries to feast on grain, a food source they ignore at other times of year. Then, and maybe only then, we can empathise with the farmer.

It is time we valued Rooks, before it becomes too late. If you're not convinced, then read Mark Cocker's superb paean to the Rook and all things corvid in *Crow Country* (2007). Lose the Rook, and we lose the magic of the period between winter and spring, when the song-dream gathers. And as for the most atmospheric rookery in England? That has to be the wild-wind-a-weaving rookery in tall drawn-up and storm-tossed Ash poles at Malham, high up in the Yorkshire Dales.

Salisbury Plain, early March.

There, where the rusty iron lies,
The rooks are cawing all the day.
Perhaps no man, until he dies,
Will understand them, what they say.

The evening makes the sky like clay.
The slow wind waits for night to rise.
The world is half content. But they
Still trouble all the trees with cries,

That know, and cannot put away,
The yearning to the soul that flies
From day to night, from night to day.

Charles Hamilton Sorley - *Rooks* (1895-1915)

Lenten Musings

The world is a mirror of infinite beauty, yet no man sees it.
It is a Temple of Majesty, yet no man regards it.
Thomas Traherne - *Centuries of Meditations*

Somehow the Western churches have managed to get away with extending Lent from 40 to, usually, 46 days – by excluding Sundays, when the faithful are obliged to be good anyway. The more half-heartedly faithful may feel that they're being ripped off, or worse, exploited. No one dare point out that Jesus only did 40 days and 40 nights in the desert, not 46.

Whatever the inflictions of the churches, Lent is perhaps the most magical time of year – bringing with it the gradual birth of spring. It is a time of deep yearning, a spring-like faith, for, to those of us who develop it – it is a yearning that gradually becomes real, as the days lengthen steadily, brighten, a green ripple slowly stretching across the land. And Lent is a time of relenting, as winter gradually relaxes its grip.

There is, though, a push and pull relationship between winter and spring: one day seems spring-like and the birds sing, the next the wind turns to the north and winter reasserts its grip. Quite often the British spring is a matter of two steps forward, one step back. There can be only one eventual winner in this cosmic tug-of-war game, though the game can be very hard fought; but the slower the spring, the greater – and deeper – the yearning, and the greater the subsequent and inevitable release.

Our Lenten spring is a time of re-acquaintance with old friends, many of which lie half-forgotten until they suddenly reappear within spring's sequence. How often do we forget that there is a patch of Grape Hyacinths or a great-aunt's favourite Crown Imperial outside the old summerhouse, until it suddenly reappears? How many of spring's lesser flowers lie entirely forgotten until they suddenly reappear on hedge bank, at streamside, or in copse – like Greater Stitchwort, Barren Strawberry or even Wood Anemone? How many summer birds fade from our minds until they return, and are welcomed – perhaps the Common Whitethroat? How many of us forget the very existence of butterflies until one suddenly appears, a lost friend, as spring lifts off?

Spring does that to us – it reacquaints and reunites us with half-forgotten friends we may not have consciously missed. And Lent plays tricks on us, deliberately in all probability, like extending itself to 46 days.

Above all, our spring is a living miracle, a time of radical transformation, transmogrification and, more obviously, resurrection that utterly eclipses the metamorphic achievements of any caterpillar. It is a deep metaphor for something almost unfathomable, quite ineffable; and Lent exacerbates it all, deliberately.

Lent can be seen as a master of ceremonies, and spring and Lent as partners in beatification. The very words and concepts associated with Lent – like Lenten, Lenten days and Lenten ways – offer a sense of deeper, truer belonging, a vital attachment to a rich cultural and environmental past in danger of banishment into nemesis as materialism strengthens its hold on us; the Lenten words hint at a richer past, and future.

Lent brings an element of discipline to spring, and perhaps more importantly, an element of deep contemplation; without it, spring might degenerate into a Bacchanalian riot, such that its deeper meanings might be overlooked, and the whole experience taken for granted. Maybe then we need Lent, for Lent is the genesis of spring.

Culkerton, Gloucestershire, early March.

I thought, they're right
it *is* the time to start anew,
and I felt biblical for that one brief moment.

Nadia Kingsley - *The day after*

Brimstone Lift-off!

All the thrushes of England sang at that hour...

Edward Thomas - *In Pursuit of Spring*

Each early spring, usually at the start of March, there is a magical hour when Brimstone butterflies take to the air after hibernating, having been comatose for six long months. You can see it coming, for an hour earlier the bumblebee queens will lift off on mass, often accompanied by a whirl of post-hibernated ladybirds, and buzz off sunwards. These insects have a lower temperature threshold than the Brimstone, which needs hot sun and a shade temperature of around 13°C.

Whatever mistakes you make in life, do not miss out on National Brimstone Lift-off Hour!

Then the butter-coloured males – perhaps the original butter-coloured flies, hence the name 'butterfly' – suddenly waken and become active. The females seem to lie abed for several days longer.

Most Brimstones seem to hibernate in Bramble clumps or Ivy clumps on tree trunks or walls. The Brambles are often sparse and straggly, not the impenetrable entanglements you might expect. Our knowledge here is, however, incipient, and based rather on assumption, for very few Brimstones have ever been found in hibernation, let alone followed through. Many records come from individual butterflies disturbed from hibernation by conservation working parties out clearing scrub or coppicing Hazel, or from being roused from their slumbers by the warmth of bonfires. There are a few records of Brimstones hibernating in tussocks of coarse grass and Holly trees.

I have followed two in hibernation: a female low down in a young Bramble patch in a sunny glade in Savernake Forest in Wiltshire, and a male high in a Holly bush in a sheltered woodland valley in the Cotswolds. Both were spotted entering hibernation on what proved to be the last sunny hour of autumn – I followed the butterflies as the weather started to cloud up. They selected leaf undersides, and hung there, leaf-like. The female became detached from her Bramble leaf after deer browsed off much of the Bramble patch. She was lucky not to have been (accidentally) eaten. She ended up spending six weeks lying flat on a bed of fallen Oak leaves, at one stage being buried by several centimetres of snow. She survived and was seen flying half a mile away in April – I had marked her wings, carefully, with an indelible felt tip pen. The male moved during a warm day in January, and was not seen again.

Brimstones seem to have favoured over-wintering grounds, often along the foot of south-facing slopes of sheltered wooded combes some distance from the butterfly's breeding grounds. They gravitate towards these places after feeding up on nectar in late summer – not necessarily close to their wintering grounds. Crucially, they enter hibernation unmated, and con-duct their courtship, mating and egg laying in spring, as do autumn Commas, Peacocks and Small Tortoiseshells.

On the first day of warm spring sunshine, Brimstone males wake up and almost immediately begin to wander, though at first they are active only for a couple of hours, from midday when the sun is strongest, before bedding down again. They are interested in warmth, nectar, and females. Nectar comes in the form of whatever's available – Daisy, Dandelions, Primrose, whatever. The pale isabelline-coloured females are ardently sought, but not necessarily on the wing, for the lemon-yellow males seek out hibernating or rousing females amongst Bramble and Ivy tangles. As I have never seen an ardent male actually find a still-abed or slow-wakening female, I am not sure how successful a tactic this actually is; but do not under-estimate them, they may well know where the females are.

This butterfly is renowned for its graceful courtship flight – the dance of the Brimstones – in which the male gyrates around a fluttering female, wooing and pursuing her with impressive ardour, in a courtship flight which can last several minutes. Such events seldom, if ever, result in actual mating for females behaving so are invariably already mated – it's just that the male refuses to take "No, I'm washing my hair tonight" for an answer and persists, until dismissed by "But I'm Already Mated" pheromones. Receptive, unmated females welcome their suitor with peaceful ease, with pairing taking place after a minimum of introduction and discussion.

Before and after mating, both sexes wander far and wide in spring and are seen a long way from any of the Buckthorn bushes on which the butterfly breeds. The poor females spend ages seeking out these bushes, and are often seen fluttering in a confused state around the wrong shrubs, notably Dogwood. They must find life extremely frustrating at times.

It must be a trick of light, for somehow Brimstone males appear more golden in early spring sunshine than during late summer.

Three Groves Wood, Chalford, Gloucestershire, mid-March.

"Yellowbird" some call you, tilting your wings
at right angles to the best of the sun -
no talk of dying out for you, my friend.
Original butter-coloured "fly",
heartening as a first primrose
glimpsed along the hard-edged way.

Keith Chandler - *Brimstone*

Primrose Picking

Children were picking primroses from both sides of the
hedges, watched silently and steadfastly by a baby in
a perambulator, not less happy in the sun than they.

Edward Thomas - *In Pursuit of Spring*

The classic Ladybird Book *What to Look for in Spring*, written
by ace naturalist and nature mystic EL Grant Watson and
illustrated wondrously by Royal Academy artist CF Tunnicliffe,
explains that there are two types of Primrose flower: one with
long stamens and a short pistil, the other with a long pistil
which is flush with the petals and calyx. The assumption, back
in 1961 when the book was published, was that children
would readily seek these subtle flower forms out, and carry
them to the school nature table, thereby ensuring that this
became standard knowledge in schools. *What to Look for in
Spring* also states that in addition to the usual five-petal form,
it is quite possible to find individual Primrose flowers bearing
anything up to a dozen petals. Inspired, I diligently sought out
these multi-petalled varieties. Over half a century later, I am
still looking, having found precisely none.

The Primrose was an integral part of rural childhood, within
the lost idyll countryside that those Ladybird Books depict, for
Primrose colonies occurred quite frequently where the soil was
not too waterlogged, sandy or acidic. Many hedge bottoms
offered yellow swathes of them in April, for this was before the
time when artificial fertilizers simplified the hedgerow flora by

promoting nitrogen-loving Stinging Nettles and Cleavers at the expense of just about everything else. Many field and stream banks flushed yellow with Primrose flowers, at least those not heavily grazed by sheep. Most copses too, offered colonies of Primroses. People used to gather them, for they were that common.

In the sleepy seaside resort of Woolacombe in North Devon, the village schoolteacher, Brown Owl and Sunday school teacher, Miss Alice Trebble (1906-1982), used to assemble the village children each Easter Saturday morning, arm them with baskets, ginger beer, and sandwiches, and send them out into the fields to gather Primroses (and wild Daffodils) in order to decorate the church for Easter Day. I know, for I was one of those children, as Alice Trebble was my Fairy Godmother. She was a veritable Maria von Trapp figure – forever laughing, eternally in love with life, childhood, and with spring. Her favourite flower was the Primrose and she would say that she was merely continuing an old country practice. She never married, but she mothered, auntied or big-sistered every child fortunate enough to come within her reach, and taught them much about natural history.

She made that church bright yellow with Primroses, set in small jam jars, Marmite jars and Shippam's paste pots, hidden amongst mounds of deep-green moss gathered from stream sides. The heady scent was almost overpowering. Years later, my sister and I went to plant Primroses on Alice's grave, in Mortehoe cemetery, only to find that the plant had already colonised the spot, naturally. We planted ours anyway, for one can never have too many Primroses, and some people deserve lots.

Today, the fields that saw these activities are all but bereft of Primroses, the unscheduled victims of high stocking rates of sheep and the side effects of artificial fertilizers and chemical sprays. But colonies persist deep in the wooded stream valleys, along the edges of Gorse breaks, and out on the broad slope of Woolacombe Down – a great whaleback hill that towers above the village's renowned three-mile-long beach. The Down is lightly grazed by cattle, and their trampling action promotes the bare ground pockets essential for Primrose propagation.

It is highly unlikely that village children will ever pick Primroses for Easter again, for the practice of gathering – sustainably harvesting – wild flowers has become not just frowned upon, but seriously scowled down. Society seems to consider it almost as bestial as Otter hunting. It is hard to determine precisely, but the causes of this major cultural shift may lie in the complexities of wildlife legislation and the messages promoted by nature conservation bodies. *The Conservation of Wild Creatures and Wild Plants Act* of 1975 gave protected species status to a small number of rare and rather glamorous plants. These were not the rarest, most rapidly declining, or most vulnerable – least of all to picking. Most people had never heard of any of them. But the Act pricked the public conscience. Then the *Wildlife and Countryside Act* of 1981, which came into force at a time of massive habitat destruction and species decline, offered full protection to a greater number of rare plants and made it an offence to dig up any wild plant. Just over a hundred flowering plants are currently afforded this 'full protection', plus a few rare mosses, liverworts, lichens and fungi. Perhaps picking and digging up became confused?

The picking of 'common' wild flowers for non-commercial purposes is still not against any law, other than on Sites of Special Scientific Interest (where legal address might prove difficult), and also on land covered by the bylaws of organisations like the Forestry Commission and the National Trust. Commercial picking without landowner consent is against the law, as it is legislated against by the *Theft Act* of 1968.   But it is agriculture and urbanisation, and to a lesser extent commercial forestry, which has led to the decline of the Primrose – not picking, commercial or otherwise.

For those who want to know, perhaps to practice this now-darkened art in the legal and ethical sanctity of their own gardens, Primroses are best picked when in the loose yellow bud stage; one or two from each clump, only – the plants flower freely over a period of a few weeks and will produce more flowers after a light picking, as they do after being grazed. Placed in a vase out of direct sun and away from radiator or fireside heat, the flowers will last a good week, whereas those picked when fully open will last but a day or two.

But nowadays, one has to wander off the beaten track to find good Primrose sites. Devon remains the premier Primrose county, for the blooms there are unusually large, and the warm moist climate suits the plant admirably. Dorset is another prime county. The Primrose is mainly a plant of early successional stage habitats, often abounding for a while where scrub has been cleared on nature reserves or where broad-leaved trees have been felled in woods; but it also does surprisingly well in pasture-woodlands which are lightly grazed by cattle, and in some old fashioned hay meadows.

But everywhere it is no longer an April flower, as it was in *What to Look for in Spring*. In the mild winters of modern times its blooms often appear in January, some even in December, and nowadays the flowers are at their most profuse during March. In all but the far north and most windward sea cliff slopes, Primrose blooms are now gone by May.

Culkerton, Gloucestershire, late March.

I wish I were a primrose,
A bright yellow primrose blowing in the spring!
The stooping boughs above me,
The wandering bee to love me,
The fern and moss to creep across,
And the elm-tree for our king!

William Allingham - *Wishing. A Child's Song.*

On Bird Song

There is a charm in some sounds so great that we love them
from the first time of hearing, when they are without associations
… There is such a thing as genius in nature.

WH Hudson - *Adventures Among Birds*

There was a time, which lasted centuries if not millennia, when many country folk could imitate a range of birds with considerable mastery. Though perhaps not quite to the standards of the entertainer Percy Edwards (1908-1996), whose repertoire was said to include well over a hundred species of British bird and whose rural sound imitations punctuated many a BBC radio programme, including *The Archers*.

I have not heard a Cuckoo imitation in years, at least not a practised and plausible one, but am aware that until quite recently the countryside resonated with these calls. A good Cuckoo imitation must include the triple and quadruple notes uttered when male meets female or when rival males meet, though not necessarily the bubble call of a receptive or egg-laying female.

Bird song imitation is a disappearing art, in urgent need of reinvigoration. The place to start a national comeback would be the Birdfair, the British Birdwatching Fair, which is staged at Rutland Water annually in late August. It would be especially interesting to hear what repertoires our TV celebrity naturalists have to offer. This should not be staged as a fun contest, but held in deadly earnest – for losing our ability to mimic bird song must represent a heavy nail in the coffin of our relationship with nature and our famous naturalists must show leadership. Chiffchaff and Curlew are relatively easy to mimic, also the simple plaintive song of the Bullfinch. I can also offer a recognisable Willow Warbler, a male Tawny Owl and an impressive Nightjar, plus snatches of Blackbird, Collared Dove, Song Thrush and Nightingale. At times, the latter has roused quiescent males into song. Unfortunately, I have rather lost my ability to Cuckoo and to imitate the Wood Pigeon, perhaps through lack of practice.

But, as in *Paradise Lost*, pride can come before a fall. I met my own comeuppance by courtesy of Jason Singh, a beat boxer and vocal sculptor from Tower Hamlets, when demonstrating my modest bird-song repertoire at an event run by Fermyn Woods Contemporary Art in Northamptonshire. Jason can imitate anything, instantly. In sporting parlance, I lost to him by a rugby score and am now officially, but not privately, silenced.

But Oh! to translate bird song into English! Is it more than, "I am fit, healthy and pumped up with avian testosterone – and this is my patch"? Is the cock Blackbird, profiled against the setting sun on the telegraph post at the bottom of your garden, merely telling other cock Blackbirds to bugger off in no uncertain terms? Is the electric storm that is the vibrancy of the Nightingale, the lord of all song, a vindictive diatribe grossly inappropriate for translation, even in a book of as base a tone as this? Does the fluting song of the Mistle Thrush effectively begin with the F word, and deteriorate from there?

Or are these offerings hymns, perhaps to a pantheistic deity that would not be out of place in Kenneth Grahame's *The Wind in the Willows*? Or are they even more than that? Is the dawn chorus an ecstatic hymn of praise, which joyously and rampantly eclipses the morning prayers our religious establishments offer?

The more I observe and listen the more I lean towards this belief, as a growing conviction, or even as a realisation. Perhaps Kenneth Grahame was on to something big, nay massive, in the chapter entitled 'The Piper at the Gates of Dawn'? Here, the book's main heroes Mole and Ratty row up river, by night, in search of a missing baby Otter. They are drawn by a mysterious music, akin to pan pipes but heard spiritually. For a long while only the aesthetic Ratty, an aspiring poet, can hear it. Nemesis is part of the music's magic, for on finding the missing child sleeping peacefully in the arms of the God Pan, the two friends almost forget the entire experience – yet remain aware that they had experienced something profoundly significant, something beyond words, which can never be discussed. When they reach the magical river island, where the sleeping babe is handed over to them by the horned God, the entranced Ratty remarks ecstatically, "This is the place of my song-dream", explaining that it is the place the music called him to. Now, we are starting to get somewhere: perhaps bird song is primarily a deep expression of love of place, a spiritual linking of soul and place through music? In which case, we humans may have much to learn, especially scientists.

The concept of the song-dream, which links living beings to their place of dwelling and time of being, and unifies them all – pauper and king, predator and prey, host and parasite, herb and herbivore – needs thinking through, especially in spring, which unifies all living things.

Noar Hill, Selborne, Hampshire, 3rd April.

"Hope" is the thing with feathers -
That perches in the soul -
And sings the tune without the words -
And never stops - at all -

And sweetest - in the Gale - is heard -
And sore must be the storm -
That could abash the little Bird
That kept so many warm -

I've heard it in the chillest land -
And on the strangest Sea -
Yet - never - in Extremity,
It asked a crumb - of me.

Emily Dickinson - *"Hope" is the thing with feathers*

The Golden Triangle

Let me sometimes dance
With you,
Or climb
Or stand perchance
In ecstasy,
Fixed and free
In a rhyme,
As poets do.

Edward Thomas - *Words*

West of the River Severn, and the seagull city of Gloucester, rests an area of gently undulating, red-clay farmland once renowned for its wild Daffodil *Narcissus pseudonarcissus pseudonarcissus* populations, and consequently known as the Golden Triangle. It is not a triangle, and never was, but a wavering corridor along the broad valley of the River Leadon.

During the late Victorian and Edwardian ages, when a branch line connected the peaceful Herefordshire market town of Ledbury with Gloucester and the great beyond, a major local industry existed, based upon the Golden Triangle's prolific Daffodils. Cut flowers were sent up to London and, more remarkably and profitably, the foliage was harvested and pressed to make a hessian-type material, used primarily for sacking. Originally a dye was extracted, but this venture seems to have been short lived. Local people, including women and children, worked in these natural Daffodil fields. They were bolstered by a sizeable force of migrant labour. Later, primarily during the 1930s, the Great Western Railway company ran special day excursions, billed as Daffodil Specials. The branch line closed to passengers in 1959, and then to freight in 1964.

Just before the First World War – which we can call by its rightful name – The Great War – a community of poets sprung up around Dymock, in the heart of Daffodil country. First to colonise was Lascelles Abercrombie (1881-1938), whose sister and brother-in-law were close friends of the local landowner, Lord Beauchamp. With a name like Lascelles Abercrombie there are only two things one can do – write poetry, or become a professor. Abercrombie did both, finishing up in Oxford. His friend Wilfred Gibson (1878-1962), the most popular and best known of the Georgian poets of the pre-1914 era, also moved there. Gibson was labelled 'the people's poet', though his fame did not last. Indeed, his writings rather predeceased him. This is a shocking thing to befall a poet, as poets, like spring and summer, seek nothing less than immortality.

Cottage rents were affordable then around Dymock, even for impoverished poets – and poets are almost by definition impoverished. Vital rail links bonded them to the poetic hub of London. The landscape basked in the Edwardian rural idyll, framed by the ancient rocks of the Malvern Hills to the north, the dome of May Hill to the south, the Cotswold escarpment out across the Severn, and a dreamland undulating away westwards into Herefordshire and distantly into Wales, the Land of Song.

The Dymock Poets, as they are now known and marketed, consisted of Abercrombie, Gibson, poet-dramatist John Drinkwater (1882-1937) and – for megafauna – Robert Frost and, briefly, Edward Thomas. It was here in the summer of 1914 that Frost persuaded Thomas, his close friend, to cross the Rubicon and become a poet, rather than remain as a literary critic and writer of poetic prose. Other poets to visit include Rupert Brooke and Ivor Gurney, the latter a Gloucester man. Prose writers were also drawn to this circle, most notably W H Davies, the renowned super-tramp, Eleanor Farjeon (of *Morning Has Broken* fame) and Arthur Ransome. Also associated with the area is John Masefield, who was born in Ledbury and in 1913 published a long story-poem *The Daffodil Fields*, which is part-based in this district. However, Masefield was not connected with the Dymock Poets.

It was not entirely paradise. The houses were cold, damp, cramped and primitive – and war came and shattered both the rural idyll and the poets' livelihoods. Two of the greatest of them perished in the war – Brooke and Thomas, and Gurney was so traumatised, the victim of 'deferred shell shock', that his mental health deteriorated and led to an early death. These are three leading lights of what is arguably the most gifted generation of poets this nation has produced.

The poets were not nature poets, with the notable exception of Thomas, but at Dymock they became nature poets – because nature did what it does to poets: it reached out to them and claimed them as its spokespeople. The Daffodils reached out to them, and entered their verse. In a rambling poem called *Ryton Firs*, which might perhaps have worked better as a prose piece, Abercrombie writes:

> … From Marcle way,
> From Dymock, Kempley, Newent, Bromesberrow,
> Redmarley, all the meadowland daffodils seem
> Running in golden tides to Ryton Firs, …

Now I breathe you again, my woods of Ryton:
Not only golden with your daffodil-fires
Lying in pools on the loose dusky ground
Beneath the larches, tumbling in broad rivers
Down sloping grass under the cherry trees

He continues, almost in Wordsworthian vein:

Follow my heart, my dancing feet,
Dance as blithe as my heart can beat.
Only can dancing understand
What a heavenly way we pass
Treading the green and golden land,
Daffodils and grass.

And reaches a descriptive crescendo with the lines:

And all the miles and miles of meadowland
The spring makes golden ways,
Lead here, for here the gold
Grows brightest for our eyes

Gibson, in a poem entitled *Daffodils*, describes:

Drift upon drift those long-dead daffodils
Against the far green of the Malvern hills,

In another poem, a trench poem called *Before Action*, Gibson remembers, 'A dream of daffodils that blow' before going over the top.

In *The Broken Gate*, looking back with melancholy and longing on an idyll destroyed by world war, Drinkwater recalls:

> ... an orchard hung
> With heavy-laden boughs that spill
> Their brown and yellow fruit among
> The withered stems of daffodil

Drinkwater also composed a poem *Daffodils* which celebrates the cut daffodil trade. It is probably not his most successful poem. Finally, Masefield describes, 'a field where daffodils were thick when years were young' (*The Daffodil Fields*).

Sadly, Edward Thomas did not live to better these offerings (his favourite spring flower was actually the Lesser Celandine, as indeed was William Wordsworth's). He did not visit Dymock during the Daffodil season. At the age of 39 his life was taken away by war, a Daffodil trampled then tossed into the muddied River Leadon.

The Daffodils became a victim of war too, but of the Second World War and its *Dig for Victory* campaign, and the massive agricultural intensification that followed, and continues, almost unabated. Today, the Golden Triangle offers but golden fragments, mere vestiges of its former regal self. Although the district remains relatively intact as a rural landscape – the M50 blunders through its heartland – the wild Daffodils, and their attendant bumble bee pollinators, have become restricted to tiny patches of relic habitat, the relics of yesteryear.

The district now attracts Daffodil tourists for a six week season in early spring. St Mary's church in Dymock and the village hall in Kempley offer excellent afternoon teas. It is clear that the plant is now integral to local identity and culture, and to the sense of spirit of place that presides patchily along the valley. But much of the farmland could be *anywhere*, as a great many fields have been rendered devoid of local distinctiveness. There are vast swathes of intensive arable farmland, offering the ubiquitous modern countryside fare of winter wheat, spring barley and hateful oilseed rape. These are intermingled with heavily-stocked sheep farms. The worst offenders are the blocks of fields under forage maize production, many of which dump what is incorrectly called silt but is actually rich topsoil directly into the River Leadon. There is also that most modern of rural scourges, an empty expanse of lifeless intensive horsey-culture.

The woods too have not escaped. The Forestry Commission's interpretation panel at the main entrance to Dymock Woods states that the woods grow some of the best Sessile Oak timber in the country – which rather begs the question of why have they been extensively coniferised? Even Abercrombie's beloved Ryton Firs, a double knoll rising high above the Leadon, has not escaped, bearing rank after sullied rank of non-native conifers. Many of the old coppiced Alder trees that line the Leadon are dying, presumably of *Phytophthora* disease. One ends up feeling like King Lear, 'Oh, I have ta'en too little care of this!'

But by modern standards the Golden Triangle is still a rural idyll – it's just that the standards have slipped, by orders of magnitude, almost everywhere. The district's Daffodils are restricted to the steeper banks that still support remnants of a meadowland flora, relics of the myriad orchards that used to adorn the district (at least, those that have not been abused by ride-on mowers), some churchyards (ditto), field hedges which have not been subjected to years of sheep abuse, some wood edges (away from the savage conifer plantations) and, crucially, lane banks and verges, with their adjoining hedges.

Yet there is still hope, and not simply because there is always hope. The Gloucestershire Wildlife Trust has established a series of Daffodil reserves, in woods, on banks and in orchards. Most of these are tiny relics; like the delightfully-named Gwen and Vera's Fields on the edge of Greenaway's Wood, named after two lady donors, and Ketford Bank above the wandering River Leadon east of Dymock. Better still, Vell Mill Meadow is an ambitious riverside orchard restoration project, complete with recovering Daffodil population and an artificial Otter holt under riverside Alders. There is evidence, albeit patchy, that the Daffodils are returning outside the precious nature reserves. These features of recolonization, and the Daffodil tourism, show that some people care, even if few people know what the tourists are actually seeking. A critical mass may well be developing. Perhaps it is beauty and love of place, and the senses of identity and belonging that go with it, that best connects the majority of people to nature, rather than rare wildlife? Iconic wildlife species only win a few people over, but beauty and place go deeper; get the relationship between beauty and place right and the wildlife should flourish.

Crucially, the issue of valuing place and landscape goes way beyond the ability of the land to earn us an income, but as a culture we are a long way from understanding and accepting this. Our growing obsession with economic issues is leading us further astray, by destroying our relationship with landscape, place, and nature. Love of place, and belonging, does not recognise ownership boundaries – and neither does nature – they offer a different paradigm. Perhaps the incipient return of the Daffodils to the Golden Triangle is linked to the development of an enhanced understanding of people's relationship with place?

My own expedition to the Golden Triangle reached its climax at St Mary's church, at Kempley, where a vortex of Jackdaws circled high above within an azure sky, celebrating Daffodil time. There, as the day's umpteenth hedgerow Robin sang of yesteryear and yesterday's tomorrow, I realised that Daffodils do not nod in the breeze, they waver. It is important to get these things right.

Gloucestershire, 17th March.

For oft, when on my couch I lie
In vacant or in pensive mood,
They flash upon that inward eye
Which is the bliss of solitude;
And then my heart with pleasure fills,
And dances with the daffodils.

William Wordsworth - *I Wandered Lonely as a Cloud*

The Corvids of Avebury

There must have been a score of them,
all lift-and-dip-and-fluster, beak and feather
against half-formed clouds, grey-bonnet winds.

The circled stones lay under them unnoticed,
and none of our paths relevant to their high view.
They flew, it seemed, without direction. But they flew.

Jo Bell - *Rooks over Avebury*

The village of Avebury in north Wiltshire gives the impression of being a place peacefully at ease, where the hurly-burly of modern times is gloriously bypassed. But places often flatter to deceive – much of the Avebury World Heritage Site's Neolithic archaeology has been deep ploughed, sacrificed beneath crops of barley, wheat and oilseed rape. Nature suffers too: few Brown Hares run in the fields, a paltry number of Skylarks rise above them, and no blood-red Poppies flame amongst the corn.

Worse has happened, for during the 17th and 18th centuries many of the sarsen stones that make up the famous stone circle were deliberately smashed up by stonebreakers, often after church on Sundays. The number of upright stones in the great circle declined from thirty-one in 1663 to ten by 1819. Then, some bright spark reassembled them.

The main road twists through the village, running slap bang through Avebury Henge and stone circle – white-van man, small refrigerated lorry, horse box, caravanette and all. Perhaps, instead of building The Dome as the Millennium Project, we could have bypassed Avebury? The Dutch would have. Sure, the traffic crawls, and people scurrying between Swindon and Devizes are able to glimpse the menhirs before dropping down to second gear to navigate another acute corner. But if you stop off and look at the stones, closely and properly, you will marvel at the colours that grow on these ancient monoliths. For these are not lifeless stones at all but intense lichen communities in varied but subtle hues of buff, yellow, grey-green and Pagan darkness, scrawling out their own hieroglyphs in a language we cannot even start to understand. A recent survey found 32 species of lichen living on the Avebury stones including one national rarity – *Buellia saxorum*, a specialist of southern acidic rocks such as the sarsen stones, of which the Avebury henge is composed.

If in spring you park up in the National Trust car park at Avebury's southern end and wander into the village, the bulk of which lies along a quiet cul-de-sac running at a tangent from the main road, peace descends. But look up, above the Lesser Celandines that line the muddied path, and there are rookeries, several and varied, scattered in Horse Chestnut and Lime trees about the village, and beyond the village boundaries in Ash groves and Beech shelter belts. The environs of the parish church of St James is a good vantage point for rooking. This is a classic English church and churchyard, complete with Saxon and Norman features, a 15th-century tower, a Victorian lychgate, a church walk – and a rookery. The church is kept open.

But the Rooks are not alone. Avebury is run by a corvid coalition of Rook, Jackdaw and the odd Carrion Crow. The Jackdaws may well be masterminding the whole show. They peer down at you from almost every chimney, and also occupy holes in trees. In winter they roost communally with their cousins, the Rooks. When the *Jackdaw of Rheims* stole the Cardinal's ring, in the Reverend Richard Harris Bareham's classic early-Victorian doggerel poem, he must have brought it here, to Avebury. Later, the Jackdaw was granted plenary absolution and canonised by the name of Jem Crow.

At the heart of Avebury stands the old tithe barn, 100 foot long with 30-foot rafters, a Grade I listed building which houses the museum that tells the story of the history of the Avebury stone circle and avenue. An imposing timber building, blackened and thatched, it is the centrepiece of the World Heritage Site experience. It also harbours an important bat roost – five species, notably a sizeable colony of Serotine bats. Outside lies a small pond in which Great Crested Newts breed in spring.

In 2013 two-thirds of the thatch was caringly replaced by Wiltshire master thatcher Ed Coney, using combed wheat reed, at a cost of some £100,000. Perhaps the work inadvertently upset the Jackdaws – perhaps because they wanted Norfolk reed thatch instead – for six months later some of them started pulling the new thatch to bits, randomly pulling out straws and idly discarding them, often in full view of the visiting public. Thatchers, ornithologists, and even corvid PhD experts were consulted, but no rhyme or reason could be found for the birds' behaviour. There was no obvious food in the thatch and few, if any, straws were being used for nesting material as Jackdaws build primarily with twigs. Theories rampaged across Wiltshire: perhaps the straws held a small starch nodule, or the avian equivalent of a legal high – ergot perhaps, or even a Jackdaw equivalent of Viagra?

Various deterrents were deployed – wooden and plastic owls and falcons, spinning decoys from allotments. All worked for a day or two only. The new thatch was even sprayed with something nasty, then double-netted with a fine mesh. All was to no avail.

Shooting was grudgingly examined, as a distant option. At this point one of the National Trust's expert naturalist advisors became particularly unhelpful, stating that he was on the side of the Jackdaws, and adding that a license would be required from Natural England, which might only become available if the protected bat roost came under threat from Jackdaw dem- olition. Shooting would not have been practical anyway, for the building and its environs are in constant use during the day and the culprit birds were only active in broad daylight, not during the quiet crepuscular hours.

Eventually an extra layer of protective mesh was stretched out several centimetres above the standard thatch-retaining wire which places the thatch beyond Jackdaw beak reach, at least for the moment, until the new wire sags. This over-netting does, though, rather resemble a hair net and may not be altogether appropriate for a Grade I listed building.

The truth is that Jackdaws take over buildings. Visit any ruined castle or one of Henry VIII's dissolved monasteries and you will find that the ruins are run by Jackdaws – from Tintern Abbey in the Wye Valley up to Fountains Abbey near Ripon in Yorkshire, across and beyond. They peer down at you from twig-filled nooks and crannies in the ancient walls, aloof and unyielding.

Perhaps it was they who dissolved the monasteries in league with the king: he wanted the dosh and a divorce, the Jackdaws wanted the buildings. Deal. But in early May the Jackdaws give us presents, for fragments of their exquisite blue and mottled egg shells can be found at the foot of the ruined walls, no two pieces looking alike.

There are times when nature simply runs rings round us, or even takes the proverbial mickey. Ask the Jackdaws, they know.

Avebury, Wiltshire, 21st March.

There were great elms in the Out-park, whose limbs
or boughs, as large as the trunk itself, came down almost to the
ground. They touched the tops of the white wild parsley;
and when sheep were lying beneath, the jackdaws
stepped from the sheep's back to the bough
and returned again. The jackdaws had their
nests in the hollow places of these elms; for the elm
as it ages becomes full of cavities.

Richard Jefferies - *The Hills and the Vale*

Up in the Wind

Spring comes when...
the lark calls louder than the crow,
the brown hedge sparrow rears and sings.
Two peacock butterflies, who drowned
in plum tree flowers, weave evening's grass
dead drunk, in charge of wings.

Alison Brackenbury - *Spring comes when*

The East Hampshire Hangers meander northwards from the South Downs on the Sussex border, up towards Selborne, before eventually petering out into Alice Holt Forest, just before Farnham in Surrey commuter country. Now within the boundaries of the South Downs National Park, the hangers consist of two wavering, serpentine lines of east-facing, steep-sloped woodland dissected by dry or damp valleys and high-lighted by treed promontories. The upper hangers occur where the chalk leads down to Upper Greensand, the lower along the transition line between the greensand and the Gault Clay. As a whole they form the western fringes of the Weald. In places the two hanger lines occur in close juxtaposition, else-where they are separated by a plateau of fertile agricultural land.

Initially the hangers appear timeless, in that they seem un-changed and unchanging. But that is illusion – for a long his-tory of coppicing rapidly becomes apparent, especially along the greensand system. The steepness has saved most of the hangers from the fate of so much of our woodland – coniferisation, for only some of the gentler southern stretches have been despoiled by non-native conifers. By and large, the hangers have escaped the ravages of 20th-century silviculture and consist of peaceful woodland, managed benignly.

The chalk hangers are clothed primarily in Beech woodland, though here, as elsewhere on limestone soils, Ash tends to be a nurse crop for Beech – for Ash follows Beech, and Beech follows Ash, in long cycles which our relatively short lives struggle to perceive. Many mature drawn-up Beech trees were blown down by the Great Storms of October 1987 and January 1990, and have been replaced by mass germination of Ash – so much so that the scars created by the storms have long since healed, adding to the illusion of stability and timelessness. Slowly, Beech will rise up through the Ash – or faster, if the now-incipient Ash Dieback disease develops extensively.

The greensand hangers are by nature Ash and Wych Elm woodland, sometimes with Field Maple and other broad-leaved trees, but Dutch Elm Disease eliminated most of the elms during the mid-1970s. Where the Gault Clay becomes prominent, Pedunculate Oak is the dominant tree. All along the greensand hanger systems there are calcareous spring lines and seepages, marked as exeats and chalybeates on the larger-scale Ordnance Survey maps.

Everywhere, in a westerly gale the wind roars uniformly and diagnostically along the upper reaches of both lines of hangers. You know you are in the hangers by the continuous roar. Many of these hanging woods possess the most poetic toponyms, as do the associated villages. The epicentre of mnemonic names occurs to the west and north-west of the railway town of Petersfield. Here lie the villages of Langrish, High Cross, Steep, Oakshott and Hawkley, and hangers with evocative names like Strawberry Hanger, Wheatham Hill, Happersnapper, Farrow Hill and Noar Hill. Even some of the lanes are glorified by names, most notably Honeycritch Lane and Cockshott Lane, which leads into Old Litten Lane.

Poets and writers are drawn to such places, perhaps as much by the toponyms – the language of place – as anything else. Sure enough, Edward Thomas lived in three houses in and above the village of Steep from November 1906 to October 1916. These houses stand above and below Ashford Hangers and the narrow finger of downland, Shoulder of Mutton Hill, which holds his commemorative stone. Here he wrote most of his poetry and worked on much of his best prose, though it must be emphasised that he spent much time away from home, especially when working on major works such as his biography of *Richard Jefferies* (1909) and the greatest of his rural prose-poems *The South Country* (1909) and *In Pursuit of Spring* (1914). Nonetheless, most of his poems were written in a free-standing, purpose-built, Arts and Crafts study room just along Cockshott Lane, at the summit of Stoner Hill. Incredibly, that building has recently been replaced by a large red brick garage.

Spring touches the East Hampshire Hangers early, often at the start of February when the Primroses on Shoulder of Mutton, and in Jack's Meadow below, begin to flower, and the Bluebell and Ramsons leaves flush deep-green along the lower slopes. The Edward Thomas Fellowship stages a series of walks on the

Sunday closest to the poet's birthday, 3rd March, come rain, shine or blow. To many of the numerous attendees, this is a pilgrimage. Periodically the strung-out groups of literary pilgrims assemble to listen to readings from Thomas's prose and verse, before moving on, ever onward and upward, for another year.

Thomas's star is in the ascendency. This is not simply because we are fascinated by the near-tragic life he lived, and his capacity for love, and despair. Beyond that, he was a man way ahead of his time. Do not think of him as a head-in-the-air poet and writer of prose-poems who lived and dreamed the Edwardian rural idyll: his time is now, for his mind was a century ahead of his time. It is not unusual for writers to be ahead of their time – DH Lawrence, for example, would have been truly in place in the 1960s – but it is hard to envisage any other writer who was actually a century ahead, as opposed to the odd decade or two. Sure, Thomas's language is distinctly yet appealingly Edwardian, as are the landscapes and wildlife he describes, burying him rather inconveniently in nostalgia – ours, not his, for he deplored nostalgia. But the messages he was putting out were way ahead of his time: we are part of nature and nature is a key part of us. His time is now in that he is needed now, acutely: his intense relationship with landscape, place and nature and, especially, his way of looking at the world and being, or rather dwelling, within it, speak volumes to us – if we deign to listen. I will come back to this later in these chapters, I hope subtly.

Today, the hangers around Steep retain much of the character that so fascinated and influenced Thomas. They collected him, and retain memories of him now. Part of him is in the wind there, in the nodding flower heads, in the Thrushes' song. Furthermore, the place, the locus, speaks through him.

The Steep hangers contain a maze of hidden paths, tenacious and slithery, so muddy that they dry out only in the most prolonged of summer heatwaves. These paths run upslope, downslope and cross-slope; many skirt the foot of the hangers, where the clay proves intractable; some run proudly along the ridge tops, the remnants of ancient ways. They pass through Beech cathedrals where boughs arch majestically overhead, through stands of tall Ash poles towering from coppice stools last cut decades ago, plunge into the shade of Yew groves that have sprung up in areas damaged by forgotten storms, and suddenly find themselves out in the open, in relic areas of chalk downland. Scarce plants grow path-side, many of them species of spring or early summer, notably that most exquisite of blue forget-me-nots, known as Wood Forget-Me-Not, *Myosotis sylvatica*, which occurs largely as a cultivar in gardens but is a genuine native, though decidedly local plant. Rare beetles and hoverflies abound, many of them associated with rotting wood, sap runs, and wood-decay fungi. And this is a stronghold of the rare Cheese Snail, a small flattened whorl snail which lives under rotting logs and in leaf litter. The hangers reek of Rarity, much of which is probably undiscovered.

But visit the hangers as April blows in: when the west wind roars evenly and uniformly along the summit trees telling of autumn gales withstood, and boasting of the resilience of hilltop trees, but when there is quietude and Hazel leaflets opening along the east-facing slope bottoms; visit then, when newly-arrived Chiffchaffs are calling silver notes from the Sallow combes and the Lady's Smock flowers, in pink delicacy, along the bottom paths and on the banks of the sunken lanes; visit when the male Brimstones are wandering, ceaselessly, spring-struck, perhaps as poets do.

Only which of the many loci, strung out like diadems along the twenty miles of hanger system, should you visit, for they are all different? Which special, solitary place is calling you? For places do call us. It is often hard to decide, for we are not in tune with the calling of a place, with the concept of the locus. Here, in the hangers, we are not dealing so much with evocative place names but with vocative places – the places that call us, perhaps so they can minister to us, or persuade us to speak for them, or at least have some form of relationship with us. Sure, you can walk the length of the hangers in a long day, by taking the Hangers Way, but then you are in a hurry, pressing ever onward, heading north or south, with little if any time for dalliance or ponderation. Yes, you can use the car, and do one stretch of hanger in the morning and another in the afternoon, after a pub lunch, which takes up the best part of the day, but then it is all too easy to have experiences – and miss out on the meaning.

So, it is often better to select just one spot – one hanger, slope, valley, combe or wood, and loiter there, intensely. But which place, which place is really calling you? Think hard on this, as Thomas did.

Ashford Hangers, Hampshire, 1ˢᵗ April.

Often I had gone this way before:
But now it seemed I never could be
And never had been anywhere else;
'Twas home; one nationality
We had, I and the birds that sang,
One memory.

They welcomed me. I had come back
That eve somehow from somewhere far:
That April mist, the chill, the calm,
Meant the same thing familiar
And pleasant to us, and strange too,
Yet with no bar.

The thrush on the oaktop in the lane
Sang his last song, or last but one;
And as he ended, on the elm
Another had but just begun
His last; they knew no more than I
The day was done.

Edward Thomas - *Home (2)*

Homage to the Mistle Thrush

This heart, some fraction of me, happily
Floats through a window even now to a tree
Down in the misting, dim-lit, quiet vale

Edward Thomas - *Beauty*

Back in April 1976, I succumbed to the dreaded mumps – in early adulthood, long past the age when one was supposed to experience this childhood inconvenience. Three days were spent lying supine in bed; any movement was difficult, if not unmentionably painful; whilst, outside, spring rose to perfection, calling me. During this time of near-delirium a cock Mistle Thrush sang almost relentlessly, favouring an Ornamental Cherry tree in triumphant blush-pink bloom close by, but ranging from perch to perch within his domain, near and far, as his kind do. Perhaps he was trying to attract a mate to his prime territory, cleverly chosen and ardently defended; or perhaps he was maintaining contact with a mate, who was spending two weeks incubating her clutch of four eggs in an untidy nest, high up in a tree fork. Perhaps the nest contained strands of agricultural bailer twine, as many do.

No way could he be shut out, for on the rare occasions when he seemed to be silent his song reverberated on in my mind. The song continued within me to the extent that I was far from sure whether it was a trick of memory, or a feature of my illness, or a genuine aspect of his song. Perhaps the song's objective was, not merely to attract or maintain contact with a mate, or safeguard a hard-won territory, but to get into the mind of other living beings, and dwell there, in conquest, even as part of spring's ministry? This is puzzling, as the Mistle Thrush song is almost impossible to conjure up in the human mind, though it is distinctive and instantly recognisable.

Thus the Mistle Thrush developed a new significance, at least to one human being. But that was no one-off experience. Other cock Mistle Thrushes have also sung on in my mind, long after my direct experience of them has ceased. There was an excellent one singing in the grounds of Birmingham University during a conference there, in March 1988 – several delegates reflected on this: he was the keynote speaker. I remember him still when passing through the University station on the train. He was definitive. There was another who sang to perfection at Denny Lodge in the New Forest during the majestic March of 1990, being part of that tract of deeply-ancient woodland itself. And so on, and onward, down the years. These alpha vocalists seek to transcend time itself.

Recently, one particularly skilful Mistle Thrush managed to flute his notes through a short section of tunnel that cuts beneath a railway line running through Hailey Wood, an ancient wood near Cirencester in the Cotswolds. The tunnel acted as a megaphone, amplifying the stentorian notes and accentuating the song, enhancing a singer who sought eternity. That bird became part of the place that is that bridge.

So, each year I tend to give out a personal Mistle Thrush of the Year award, as a gesture of thanks to one of the eight species of British songbirds I would take with me to a desert island in lieu of utterly useless and one-dimensional gramophone records. The fact that in 2015 this species was placed on the European Red List, having undergone a significant decline, should concern us all, for it is one of our primary songsters, active in both town and country. Its song is an integral part of the music of the song-dream.

Culkerton, Gloucestershire, 6th April.

At once a voice arose among
The bleak twigs overhead
In a full-hearted evensong
Of joy illimited;
An aged thrush, frail, gaunt and small,
In blast-beruffled plume,
Had chosen thus to fling his soul
Upon the growing gloom.

Thomas Hardy - *The Darkling Thrush*

Through a Glass, Darkly

For now we see through a glass, darkly…

1 Corinthians 13:12

We increasingly experience nature and the great outdoors through windows, as we hurtle through landscapes at ever-increasing speed. As speed increases so the glass – or Perspex – strengthens, and in doing so allows something dark and distancing to strengthen its grip on us. Our sense of detachment increases and normalises, so that viewing – experiencing is the wrong word – becomes a passing interest, lacking any vestige of true involvement. At best we have some petty experience, but grasp little of its meaning.

But in mid-April so much hurtles by anyway, whether we're in a car, on a train, or actually out there in the real world, the wild. The pace gathers as spring intensifies. It is a hurly-burly time. Bacchus and his wild-eyed followers run riot.

My intercity train flashed by the first leafing Oaks, the last of the puritanical Blackthorn blossom, Sallow blossom, rookeries a tree-top, fields sown with spring crops, Hares and ascending Larks, and gardens that hinted of summertime. It all looked so idyllic, even the vast acreages of oilseed rape seemed natural, for at speed and through the mindset of the screen one cannot tell true gold from false. Even a field sprayed with herbicide to kill off agricultural Black Grass deceitfully flashed an alluring burnt umber under the April sun.

The Perspex screen distorts our experience, and so confuses us. It is perhaps yet another manifestation of how distorted our relationship with Nature has become. Even the sense of distance is altered by speed and glass, so that the landscape seems almost two-dimensional. And we cannot hear the bird-song or smell the blossom, let alone breathe in the energising spring air: in a train carriage one is treated to the *Te Deum* of businessmen on mobile phones and whiffs of the acrid scent of train brakes.

In effect, the more we hurtle through landscapes, seeing transient vistas at random, the more nature becomes something to which we falsely relate; and the more we become trapped in a sanitised bubble behind the Perspex screen.

The railway cutting banks were littered with the trunks and branches of trees and bushes felled primarily to facilitate the electrification of the line between Bristol and London, but also to prevent that most catastrophic of commuter train events – namely, delays caused by leaves on the line. For nature had been slowing us down, by littering the line with fallen leaves, in autumn no less, when the nation is gearing up for that most important of economic events, Christmas. So trunks, branches, lop and top, brash and brushwood, were all being left to rot. Time to look down at one's reading material.

And on this April day people were not out in the sunshine, looking out for the Swallows that were arriving, or listening to the drone of bumblebees that tell of timeless summer days to come. Not even a child. They were all in buildings – houses, offices, places of work, examination factory schools – or they were in vehicles, hurtling through landscapes as fast as they could, or stuck behind each other in mindless traffic jams.

All save one, who was on a London-bound train and was glued to the window, drinking in as much Nature through the thick screen as he could; projecting himself out there, to the other side of the Perspex screen, to the sunshine and amongst the unfurling leaves – into the real world.

Some reaction to the false imprisonment was necessary. It began with a toddler tantrum in carriage F. The tantrum inten-sified, reaching Storm Force 10, and spread to another small child, in carriage E; and so on, until every carriage was scream-ing. The message, translated out of baby babble, was simple: we are trying to live a lifestyle we were not designed to live.

Children must lead the rebellion.

Great Western Railway to London Paddington, 13th April.

A man that looks on glass,
On it may stay his eye;
Or if he pleaseth, through it pass,
And then the heav'n espy.

George Herbert - *The Elixir*

In Pursuit of May

The hedgerows itch with last summer's nests;
unwise saplings try out bright buds on their wrists.

David Morley - *A Spring Wife*

Spring rushes and gushes from one thrill to the next, leaping ahead of itself. I decided to play that game too, and venture out in search of May – in mid-April, when April was running slow and was clearly not in the giving vein. This search is as easily done as said: just hop on a train and visit any of the central London parks, as there the seasons are always ahead of elsewhere, ahead of whatever norms there might be – so much so that it appears that nature too has joined the London rat race. It's the extra heat that does it – the heat that escapes from our over-heated buildings and underground network.

My previous visit to London had taken place in mid-December, when Regent's Park had been dressed for early February with Hazel catkins, beds of indigo Polyanthuses, and saffron Wallflowers in full bloom; Narcissi, blue and white Periwinkle, and Witch Hazel welling out; any amount of pink Viburnum and other blossom – and mating ducks. That's London for you in a mild winter. The seasons run later in the vast, empty expanses of grass – at best, Daisyland – and amongst the avenue trees.

The loveliest of the parks has to be St James's, the oldest, but any of the bushier or more flowery parts of any of the other Royal Parks will do. It is to the shrubberies and borders that most people escape, over-spilling on the hottest days into the hinterland. It is here that most benches are sited. It is here that most of the parks' wildlife can be seen, and not just the Grey Squirrels.

Sure, there are water birds aplenty on and around the lakes – a hotchpotch of resident, migrant and naturalised species, and their hybrids. The Great Crested Grebes, sadly, were not courting on the Serpentine when I visited on a sublime day in mid-April, though doubtless they were head-bobbing and giving and receiving weed love troths elsewhere that day. But in London they had long gone through that phase, and as May-like sunshine smiled down from sapphire skies they were far too busy dodging the pedalos, which were out in force on the water for the first time that year. Further proof of advancement was provided by the teenage hen Blackbird, hopping around along the edge of The Dell. She must have been laid as an egg in January and now, roaming around free of parental guidance, was fully able to forage for herself. Her parents were probably busy attempting to rear another brood; rightly so, as we cannot have too many Blackbirds in this world.

In the shrubberies near Hyde Park Corner, male Holly Blue butterflies, as blue as the sky, were following pheromone trails along the higher contours of the shrub edges. They were looking for females but mostly encountered each other – just like blokes. The Royal Parks are prime Holly Blue country. In spring, the females lay most of their eggs on the sepals of Holly flowers, but struggle to differentiate between male and female flowers. That's a major botanical blunder as the larvae fare best when they are able to bury into developing Holly berries, and the unfortunate larvae on male trees have lower survival rates.

There's an even bigger question over where the summer brood females deposit their eggs in the parks. The butterfly textbooks state that summer brood females lay their eggs exclusively on the buds of Ivy, a plant all but absent from the Royal Parks where it is regarded as a hostile foe and is diligently removed. Perhaps some bright young ecology student from Imperial College can research this anomaly. There are gaps in ecological knowledge that are probably more significant than anything we already know, but those gaps are poorly recognised.

But yes, I found May in mid-April in Hyde Park. The Tulips and Spanish Bluebells were in full bloom, whilst the Daffodils had long finished. Cerise Crab Apple blossom was at its height, only it was not attracting the profusion of bees and hoverflies one finds on Crab Apple blossom in rural woods or hedgerows. I saw but one hoverfly – an aphid-feeding species of spring woodland, and none of spring's mining bees; only an abundance of the common Hairy-footed Flower Bee *Anthophora plumipes* and the Common Bee Fly *Bombylius major*, feeding together on *Erisymum* 'Bowles's Mauve'. The juxtaposition of those two insects was interesting, for the bee fly is a parasitoid of mining bee larvae but is not supposed to be associated with flower bees: another student project in the offing, perhaps. Entomologically, the visit was of interest for what wasn't there, rather than what was; but that's inner city invertebrate faunas for you. I had found May, not just in the advancement of flowers, blossom, birds and foliage but in the strength of the sun and in the intensity of light, reflected off towering buildings, and was reassured that May would soon come everywhere, and make spring even better than it already was. I could not wait for May, and neither could London.

Hyde Park, London, 13ᵗʰ April.

Now the bright morning-star, Day's harbinger,
Comes dancing from the East, and leads with her
The flowery May, who from her green lap throws
The yellow cowslip and the pale primrose.

John Milton - *Song on May Morning*

The Spirits of the Spring

Somewhere among their folds the wind was lost

Edward Thomas - *March*

The further time takes me away from the long hot summer of 1976, that doyen of all summers, the closer I get to it; because the more it calls me, homeward. It grows in my mind not as something historical, for rigid historic facts matter decreasingly over time, but as a calling, as a desired state of being which needs to be found again: the long hot summer – a paradise lost, that needs to be regained. I have become in love with a by-gone golden age, and what it represented to me – the state of mind I was in that year; but then we are all lovers of epochs of time, naturalists especially.

The English language has provided us with a word for love of place – 'topophilia', which is attributed to the poet John Betjeman, though the word lacks poetic resonance and sounds wholly un-Betjemanesque. We also have a word for the love of biotic nature, or wildlife – 'biophilia'– but there seems to be no equivalent term for the love of a period (or periods) of time. Perusal of the Classical Greek dictionary suggests the concept of 'ainophilia' – concerning the love of memorable passages of time, both short (like a week's holiday, or even a memorable day) or long (like the love for a golden age). Ainophilia can be deeply personal, or something many people share. Always, it is hugely positive. It is derived from αἰών, meaning a period of time linked to personal destiny. Perhaps we need this word, to enable our memories to develop into understanding, and gather depth, and to clarify our relationship with time.

It was therefore a pledge, to the long hot summer rather than to myself or to God, that I would see my first Orange-tip butterfly of this latest brave new year in the place where I had encountered the first of that golden summer, four long-spent decades ago. That experience took place in the 'country' of Gilbert White, the first and foremost naturalist, globally, near Selborne in Hampshire. The day was Friday 16th April 1976, when high pressure was building from the south-west, Oak leaves were unfurling and Spring herself was stretching out before me, grass-skirted and Daisy-chained. Then, after an hour's wandering search, a freshly-emerged and weak-flying male ambled slowly past me, as if in a dream, following a stream that meandered down a secluded valley, part wooded, part marshy grassland. So, my magical encounter took place at the epicentre of natural history, where it all began.

Moreover, my first Cuckoo of the year was also met with there, that halcyon afternoon, a sleek black male with a deeply-barred chest. I disturbed him from feeding on the ground, on furry moth caterpillars in all likelihood, but froze too late, and had to watch him fly off as silent as a Sparrowhawk, into the leafing trees. Later I heard him call, once, then again, itera-tively. He had come home. To me, he was not calling out for a mate, but calling up summertime; mates would come later, when the time was right – get Time right, the rest will follow.

Back in the 1970s Orange-tips normally began to emerge in mid-April, peaked in early May and then lingered on into mid-

June. Their flight season coincided perfectly with Cuckoo-calling time. Nowadays, in southern England, Orange-tips often start to emerge in late March or early April, before the Cuckoos arrive, and are gone by the end of May. Furthermore, the Cuckoos are all but lost from the Selborne area, until recently a stronghold, and have indeed vanished from large parts of lowland England. But 1976 was part of the time before the advent of climate change, before the time when butterfly flight seasons began to move steadily forward, and before the era of mass Cuckoo-loss. That year, despite a hot sunny spring, Orange-tips persisted in East Hampshire into mid-June, when the Cuckoos were finally ceasing. The Orange-tip is, like the Cuckoo, no mere harbinger of spring, but an integral spirit of the very April air, a wanderer of sacred places in sacred time; it is an integral element of spring in England.

So, I dutifully returned, forty years on, on a day of billowing clouds punctuating a cyan-blue sky, half-fearing to find the valley despoiled by game-keeping and Private signs, for I had not walked that way for nigh on thirty years. I should have had more faith, even though faith is irrational.

I remembered little of the valley, or little of any consequence beyond its geographical position. It had changed almost beyond all recognition, having entered an era of benign neglect, a successful escape from the maelstrom world. Moss-covered trunks of collapsed Ash and Sallow stools were strewn about, lying in the stream and along the edges of open marshy areas. The valley-side woods were littered with fallen branches, which spring was busy burying beneath drifts of Bluebell and Ransoms. The sinuous meadows were no longer grazed, except by deer, and the fences had become rickety. This was a world apart, withdrawn and self-intimate. No one had been there in months, for the valley had become almost impenetrable. As I squelched along the vestiges of a footpath, over which wet winter seepages had erupted so spectacularly that the saturated path had become an aureate river of Alternate-leaved Golden Saxifrage, I sensed that something of the place remembered me, and had called me in, homeward. It was not the butterfly which had called me, but the place. The further I ventured, slipped, squelched and stumbled, the more I realised I was a lone invited guest, and that all would be well. I was entering a place of calling, where the song-dream had brought me.

And yes, first one then quickly a second of these dainty tropical-looking butterflies drifted past me, at speed, carried by their own power and not by any breeze – flame-orange banners on vestal-virgin wings, dappled with vivid green beneath for camouflage when at roost, bright forewings pulled back. They moved surprisingly fast, and constantly, for such a frail-looking butterfly; but they are much tougher than they seem, with the character necessary to survive the vagaries of English spring weather. Perhaps I had watched their grandfather thirty-eight times removed, all those years and generations ago, when the valley trees were younger and less mossy, less fallen, less withdrawn, and the whole world seemed more verdant and beckoning.

They did not follow the same route as their ancestors, for the valley had changed too greatly, but they followed the same journey, the same urge, and perhaps the same faith. They were moved by the same power, a power beyond wings and words.

But what was I seeking, and did it matter? I was not seeking mere repetition, for lost years do not come again, not even one as mighty as 1976. No, I sought to redeem time itself: I mean the wasted years, the days spent mindless in offices gazing vacuously through double-glazed panes behind which spring drifted gloriously by, on eiderdown pillow clouds and far-flung Swallow wings. I had come back to apologise for a life half-lived away from nature, and for a life hijacked, and often wasted, by the full-blown and under-achieving bureaucracy that nature conservation has sadly become.

Above all, I had come back, to give thanks – for thanksgiving is a wonderful experience – for two heady and inspirational hours spent four decades back-along, which transcended time itself by becoming immortal, at least within the confines of one human mind and, perhaps, one forgotten valley.

Then, a long leaden cumulus cloud cast its spell over the valley, and the temperature dipped back towards March. The mining bees and hoverflies descended from the Sallow blossom to bask on warm fallen Oak leaves that would soon rot down, the cast leaves of yesteryear. And I knew then that I mattered not, but that my experiences in time, season, and place did matter, when communicated and shared.

And sudden, where Brambles and last year's Bracken fronds had collapsed in, over a tumbledown fence – another unnec-essary man-made partition – a cock Wren was busy, building a nest of twisted dead Bracken fronds and moss gathered from Ash boles. He would construct more than one nest, but I sensed that this would be the one his mate would choose, and was confident that there would be Wrens to chide any stranger who deigned to set foot in this the loneliest and most fortunately neglected of valleys, in pursuit of summer.

Finally, for there was a need for finality, from somewhere down the twisting valley, with its incised stream lined with liverworts, beds of yesterday's sun-warmed Bracken litter, golden-blossomed Sallow thickets, and steeply-sloping woods with Bluebell hues, a Cuckoo called – just once, as if in passing, softly; then once again, as before, but now as if from another era, even more distantly.

Burhunt Valley, Selborne, Hampshire, 16th April.

The glory of the beauty of the morning, -
The cuckoo crying over the untouched dew;
The blackbird that has found it, and the dove
That tempts me on to something sweeter than love;

And I shall ask at the day's end once more
What beauty is, and what I can have meant
By happiness?

Edward Thomas - *The Glory*

In the Valley of Silver

Motorways grub like glaciers.

He suns. He sleepwalks on the wing
through this world and the next

Roger Garfitt - *Buzzard Soaring*

Spring was flowing north. It was time to join the red snake of
brake lights that squirms northwards through Birmingham to
the unspectacular landscape below the Pennines' solemn western flank. For once the M6 was in benevolent mood.

The Willow Warblers had arrived in Silverdale, on the
Lancashire-Cumbria border, ahead of me. Their mass pipes
were roistering in the scrublands and Birch coppices, forming
a near-continuous chain of musical diminuendos along the
labyrinthine paths that characterise the district. Overhead,
Ravens croaked, prophesising summer, and a Sparrowhawk
circled on stilling air. In the distance a chainsaw groaned and
grumbled; it needed sharpening, and sounded as though it was
protesting its way through Yew, a hard and grudging timber.
Whatever, it intruded into an idyll that otherwise seemed too
lovely to be of this age.

Silverdale understands scrub. The village itself is disparate,
spread out over a vast acreage of undulating countryside of
rock-strewn limestone hillocks, peaty valleys and hollows. But
Silverdale is far more than a village, or a parish; it is a state of

mindfulness, a place apart, and exists within its own time zone. Everywhere there are woods and copses which were formerly coppiced for tanning bark, firewood, and the great Lancashire bobbin industry – the dark Satanic mills must have got through a mighty lot of bobbins, given the scale of coppicing evident from the plethora of old coppice stools. If it grows, cut it, and cut it again; that must have been the maxim. Nowadays, coppicing is carried out by nature conservation organisations, for much of our wildlife is well adapted to the coppice cycle, following the woodcutter, and has declined considerably in abundance since coppicing declined.

The Silverdale woods stand on rocky ground, too difficult for agriculture which has tamed and shamed the easier land. Here trees root themselves into just about every crevice between the limestone joints, though their growth is unusually tardy. Ash and Silver Birch abound, but others also, and with Hazel all but ubiquitous as understorey, or underwood. The rockier places have been invaded by Yew, which perhaps owes its standing as a tree too-sacred-to-fell to the fact that its timber is so hard that no labourer in his right mind would want to chop it down. And so a bail-out myth was invented: every Yew is sacred, planted by the ancient druids, so leave it. Scrub Oak abounds, much of it growing from coppice stools, telling of the tannery industry gone by. Beech, Holly, Hornbeam and White-beams occur freely too, and the true scrubland species of Blackthorn, Hawthorn and Sallows. Almost everywhere, Ivy seeks to entwine. Underfoot, mosses grow profusely, over boulders and tree boles, wherever there is shade. Yet the ground flora is remarkably rich – indeed, this is one of the best areas for botanising in the country: I found the distinctive four-lobed foliage of Herb Paris in one broad grike, between mossy boulders.

Lanes and footpaths that twist, turn, rise and dip repeatedly, link the plethora of features that make the Arnside and Silverdale Area of Outstanding Natural Beauty so remarkable, and so intimate. Magical place names, 'toponyms', abound – like the Fairy Steps that run up a crack in the limestone scar at Beetham Fell, the nearby Creep-i'-th'-call Bridge, and Cringlebarrow Wood. The village names of Yealand Storrs, Yealand Redmayne, and Yealand Conyers speak volumes. And this is pepperpot country, for man-high cairns are dotted about almost willy-nilly and not necessarily on hill tops and promontories – there's a remarkable one in the middle of an expanse of scrub-invaded limestone pavement in Gait Barrows National Nature Reserve, though it's not easily found and even less easily found again.

The district also specialises in sunsets. Often the shallow estuarine waters gleam pure silver – hence the name Silverdale, though they flame orange at sunset time before turning to rose at dusk. On clement evenings people drink in the beer and sunset, in equal measure, from the forefront of The Albion, on nearby Arnside's Victorian promenade. It is the sunsets that rivet them, rather than the distant views of the Lakeland fells: those, they take for granted.

This is open-secret England: the masses race past, on route to the honeypots of Ambleside and Windermere, Wordsworth's beloved Grasmere and Beatrix-Potter-land at Hawkshead, leaving this district for its own faithful. And that has been the case since the time of Gilpin and the Picturesque, Wordsworth and the Romantics, and Ruskin and his movement – they all scurried past the low wooded hills of Morecambe Bay in a hurry to reach the epicentre of the Lake District. In Lakeland, it is usually raining; in Morecambe Bay, though, the climate is significantly sunnier, and drier. Often, one can watch doom and gloom descending on Lakeland from the sun-drenched hills of Arnside, and smile contentedly.

Deep history is curiously obscure here. There is but one Neolithic fort, a paltry affair above Warton at the southern end of the hills, and precious little in the way of myth and legend. Religion too, both pagan and Christian, has also failed to make much of an impression. So the district is not anchored anywhere in human time, or with any particular strong human identity, but has its own sense of time, and its own identity.

The intensity of light here is unique. There is something about the way light reflects from the grey limestone walls, paths, rocks and slopes of scree and clitter; but at the same time the trees, especially the sombre Yews, absorb light. The result is a curious juxtaposition of lightfulness and darkness. One day a great artist, writer or poet community will develop here, and change the face of Art forever. They may centre their attention on nature.

The terrain is suitable for pottering, rather than strenuous hiking. That makes it ideal second-honeymoon country, where people of a certain age, in the period of life between children leaving home and grandchildren arriving, can walk hand-in-hand without feeling embarrassed. The district is deeply revered, being the place to where every Lancastrian wishes to retire, or at least have a timeshare in a caravan. This is a land of love, adored by the people who have discovered it. Here, its lovers can reconnect with their inherent love of place and nature, and live fragments of the lives they imagined. For those who have failed to establish a pied-a-terre here, there are numerous houses, closes, and cul-de-sacs called Silverdale elsewhere in England: for Silverdale casts silver shadows elsewhere, and gilts our minds.

The small town of Arnside, on the south side of the Kent estuary, is run by summer's Swifts, only on this particular trip I was too early for them – as the main pulse of Swallows, and early warblers, was only just arriving. At dusk, skeins of geese and flotsam seagulls fly slowly down the estuary to their roosting sites. At the first flush of dawn they return, to their feeding grounds. The estuary rings with the haunting calls of Curlew, Oyster Catcher and, most plaintively, that eternal watchman of estuarine sands, the Redshank.

I woke in Arnside, to the sound of Collared Doves, and Wood Pigeons, proclaiming that the day was set fair, and that the pale haze now showing would metamorphose into pure azure. Cloudlessness was promised. In the nearest wood, a Song Thrush was making a party political broadcast from the top of a bare Ash whose flower buds were swelling. A little further uphill the first of the day's Nuthatches was trilling high on a dead branch of a veteran Oak. There was joy in the air, and on the ground too: for Violets thrive in this district, in the forms of the Hairy Violet, a denizen of limestone soils, and the Common Wood Violet – both were generating a purpling haze along the woodland paths. Southern Wood Ants, one of many species that occur here at the very northern limit of their UK distribution, were warming up together in pools of sunlight. Soon they would begin to rebuild their heaped nests of arboreal bric-à-brac, wander every millimetre of the woodland floor, and explore every bark crevice, low and high; but for now they were content to behave like normal invertebrates and squirm and writhe in the April sunshine.

From the summit of Arnside Knott, the prominent hill above the grey-stone town, mid-April lay dreamily before me, in leafing mode. A universal greening was taking place, slowly but surely, utilising an impressive range of unnamed vernal greens. Although there is another knotted hill some way to the east, Farleton Knott, Arnside's own Knott is unique, certainly within the context of England. Part open, dominated by scree and Blue Moor-grass, whose flower heads waver in any breeze, and part wooded, with groves of twisted trees and gnarls of exposed root, the Knott offers unique views, which are not mere vistas, but dreamscapes into other worlds. To the east, across an eclipsed M6, stands the grey massif of Farleton Fell, Holmepark Fell and Hutton Roof Crags, where the Carboniferous Limestone hills of Morecambe Bay meet and merge with their larger cousins in the Yorkshire Dales. But eastwards, the view from Arnside Knott is always hazy, and Yorkshire seems a world apart – cut off, by itself most probably. Nearer, is evidence of the RSPB's ministry in the district, for the nature conservation charity has been steadily acquiring Silverdale's peatland valleys, taking them out of intensive agriculture and turning them back into the peaty wetlands they once were. This redeeming work is masterminded from RSPB's reserve of reed bed and open water at Leighton Moss, on the south-east edge of Silverdale. There is a process of mending, of reintegration, here.

From the southern crest of the Knott, if one averts one's eye from a forest of poorly-concealed caravans, Silverdale lies spread out behind its silvered sea. Distantly, Warton Crag, the southernmost of the limestone hills, tells where the dream-world ends. To the near-west lies the Oak-wooded promontory of Arnside Park, and beyond the dazzle of Morecambe Bay, the retirement town of Grange-over-Sands and far-off Sandscale; further still, there is only haze and a sense of an irrelevant else-where. The Knott, and its westerly sister hill of Heathwaite, offers eternal sunsets – sunsets which are so impressive that they linger on within the minds of those who have experienced them.

Northwards, across an estuary of ever-changing sand banks and river channels, stands another range of carboniferous lime-stone hills, wilder, haughtier and loftier, and the bleak raised bogs of Meathop Moss and Foulshaw Moss; then, further still, when not lost in haze or cloud, the Lakeland High Fells. Tech-nically, one can identify each Lakeland peak from the National Trust's toposcope on the Knott's western summit; in practice, only a few are identifiable – notably Coniston Old Man, Scafell Pike and that most distinctive of all Lakeland features, the Langdale Pikes. But it is the near-distance that holds the eye most: the Knott's freestanding wind-thrown trees which grope leeward and tell of autumn gales past, stunted Yew trees topi-arised by browsing deer, prostrate kelly-green Junipers, and most of all, the isolated Scots Pines on a knoll on the upper slopes of Redhills, just down from the Knott. It is those pines that best characterise and define Arnside Knott.

The Knott is not the most tranquil of places: trains rumble noisily across the viaduct, the incoming tidal bore is announced by a Second World War air raid siren, warning bleeps drift incongruously up from the town whenever a delivery lorry reverses, and there is inevitably a ride-on mower or builder's drill in operation down below. But the Knott rises above all this and offers many moments, even hours, of quietude. Above all it is a place where a small part of your soul will remain behind, when you leave, to call you back, magnetically, and back again, as a touchtone. It is a deep heartland, a place of endless belonging, and of pilgrimage. It stands out within the wider landscape of lesser hills, appearing suddenly in gaps between other hills, beckoning, calling for its faithful to return. It is the fulcrum of this landscape.

In mid-April, on the first sunburn day of the year, the Knott was busy mending itself after another gruesomely wet winter. The Blue Moor-grass was flowering, for it is the first grass of the year to bloom; and the first of the Knott's botanical gems was coming into flower, the Teesdale Violet, a northern limestone species. And everywhere there was evidence of mending, of coming back into being, of spring's arrival and the promise of summer to come.

Here, spring was mending me, too. But my journey led me on. One cannot relax in spring, for there are too many adventures to encounter in spring's frenetic hurly-burly metamorphosis. One has to metamorphose too. And other places were calling.

Edward Thomas stayed at Silverdale late in 1916. One of his closest confidants, the poet and dramatist Gordon Bottomley had moved there from Grange, across the estuary. One of Thomas's final poems (*The Sheiling*) was inspired by his friend's house, built high up on the hill by Elizabeth Gaskell's daughters. The poem begins –

It stands alone
Up in a land of stone
All worn like ancient stairs,
A land of rocks and trees
Nourished on wind and stone.

Willowfield Hotel, Arnside, Cumbria, 21st & 22nd April.

A Flower for St George

If they had reaped their dandelions and sold
Them fairly, they could have afforded gold.

Edward Thomas - *Lob*

A slow spring is no bad thing, as a fast-track spring almost invariably becomes unstuck – the jet stream jumps south, maliciously, at the end of April or start of May, heralding six weeks of dismal weather. When spring capitulates like that its winged insects are quickly written off, having overreached themselves, and much of the nesting season is desecrated. We have suffered several such springs in recent years and are over-due some (considerable) relief.

St George's Day is the fulcrum of spring, from which the season should leap into full leafing and total bird song, and when the dragon of winter is ceremoniously slain by a knight in effulgent armour. We subsume the day within myth and legend: Morris dancers issue forth, for the first time in the year, and silly things happen in village pubs. More purposefully, the St George's flag flies from church towers, for this is one of the few occasions in the year when Englishness is actually celebrated. It is not so much a sense of national pride that is lacking here, as a sense of belonging – rootedness – and identity. Perhaps English families have moved around so much since the indus-trial revolution that people have become shallow-rooted? Spring should help connect people with place through nature, and connect people with nature through place – especially on

St George's Day, when time, place and nature can blend together, weather-permitting.

This St George's Day dawned clear but cold, borne on a northerly airstream which originated from towards the Arctic, and was clearly intent on checking spring's advance. For a week or more Dandelions had been rampant, most visibly along road verges and in old-fashioned meadows.

Formerly valued by herbalists and aesthetes as a useful plant and as a powerful emblem of spring, this flower is now at best taken for granted, even though it mimics the sun in miniature and nestles in a homemade bed of deepest vernal green. It was not valued by farmers, for a profusion of Dandelions in a meadow has long been viewed as a sign of poor farming, indicating that heavy grazing had taken place during the previous summer – though sometimes it was telling of a hot summer gone by, when grass was in short supply, and heavy grazing inevitable. Many farmers spray them off, out of embarrassment as much as dislike. Whatever, the Dandelion is a pioneer coloniser of bare ground, grows incredibly fast in early spring, and flowers profusely, reflecting spring's sunshine in all its glory.

These days, we are doing ourselves, and this flower, a disservice. We need a National Dandelion Week, Fortnight, or at the very least a National Dandelion Day, in which this most beautiful and valuable plant is openly celebrated. In years when spring arrives early, Dandelions may be at their most profuse during the second week of April but, when spring is late, peak flowering may occur during April's final week. Almost invariably, though, the plant is prominent on St

George's Day – provided it is not raining, as then the petals are closed up. At the very least, the Morris dancers should adorn themselves with its golden crowns, and gardeners graceful enough to allow the plants to flower before howking them out before seeds are set.

It is preposterous that we vilify this plant for it is one of the most beautiful, and useful, of our native herbs having culinary, medicinal, and viniculture value. It is of immense value to spring's insects as a source of nectar and pollen. Think bees, butterflies, Dandelion and Burdock, Dandelion tea and wine. Think Rabbit food. Think childhood – for Dandelion seed clocks are as essential to childhood as Daisy chains and sand-castles. Think Dandelion time, for when the plants have finished flowering their parachute seeds drift idly by, from spring into early summer. And the plant has many alluring local vernacular names, like Cankerwort, Lion's Tooth and Witch's Gowan – each of which tells a tale.

To expert botanists, our so-called 'common' Dandelion is a fascinating and highly challenging complex of some 229 micro-species, of which at least 40 are endemic. There are specialist botanists who work with the family *Taraxacum*, Dandelions, only. Dandelions are that fascinating. There are some seriously rare Dandelions too, specialists of montane ledges, coastal habitats and calcareous fens. Perhaps our attitude to Dandelions illustrates much of what is wrong with our relationship with Nature? We want to be in control of this plant, but it is too clever for us – too resilient, too adaptable and exploitative: it refuses to be controlled, and it benefits from many of the messes we make. Perhaps we need to flip this coin over, and start to appreciate and value the Dandelion, and respect it? Yes, we need a celebratory National Dandelion Week each April, centred around St George's Day – it is there, on offer, along our road verges, for all to see.

Hilton Park Services, the M6, 23rd April.

A charm of goldfinches
scatter over roadside dandelions.

Gorse-bright,
a prick of blood behind the beak,
they steal blowing silver

and spin it from down
to gold filaments.

Chris Kinsey - *A Charm of Goldfinches*

May Day Journey

Much has been written of travel, far less of the road
Edward Thomas - *The Icknield Way*

The A272 trundles eastwards from Winchester into the oaken kingdom of Sussex, where it eventually disappears – into vapour most probably. It is one of the loveliest roads left in England. Driving this undulating, meandering road is one of the few pleasurable experiences left in motoring, except on the occasions when one is stuck behind a lorry, or more likely here, a horsebox.

The richest section begins just before the road crosses from Hampshire into West Sussex, in Edward Thomas country. There, at the junction with the old A3, leans a giant Scots Pine, its crown resting upright on the ground. It was half-felled by the Great Storm of October 1987, and remains there, alive but growing at a drunken angle, a slouching sentinel of autumn, disrobed. It could be utterly ignored on a day of slanting sunshine and towering cumulus mediocris clouds as April was surrendering to May.

Our road keeps the rounded forms of the South Downs escarpment a little to the right. It is fringed by Dandelion verges and offers frequent patches of other spring flowers – Bluebell, Garlic Mustard, Greater Stitchwort, Lady's Smock and Primrose, and more Spanish Bluebells than botanical purists would care to mention.

This road is about mending, the mending of our relationship with time, which mid-spring provides, and the mending of our relationship with natural beauty, and with landscape. Journeying along it is restorative to the human mind too, traffic permitting. There are signs of mending all along. First, just into West Sussex lies Durleighmarsh Farm, a small family enterprise offering excellent asparagus and other organic vegetables, a tea barn where one can indulge in the delights of cake in a corner of quintessential England, and where the restoration of rushy meadowland is well under way. Past efforts to turn these fields into more productive pastureland have been reversed by enlightened owners, so that Orange-tips and Green-veined Whites once again dance over meadows filled with their beloved Lady's Smock flowers, and with the promise of Buttercups to come. A group of bullocks had just been turned out to grass here, after being cooped up all winter in a barn. They were cantering around the field, tails up, as happy as cattle can be. Later they would simmer down into a summer of bovine contentment.

Next door, main crop potatoes had been planted on an industrial scale. This farm doesn't do acres, it does hectares, and intensive hectares at that. I ignored it, distracted by a massive bank of Stitchwort opposite, a pair of Swallows overhead, a flowering Wild Cherry in a lay-by, and then by the soft green of fresh Field Maple leaves through which flower buds were appearing.

The village of Rogate gifted the last of the Grape Hyacinths, a Magnolia in full pomp, a Blackbird dipping down to cross the road, and a church wedding exodus complete with a vicar dressed as an leucistic bat, confetti and wedding bells that shook passing vehicles. The bride wore pink. Translated from the Latin, Rogate means Ask! (or Pray!) in the plural imperative. Years ago, it may have been in 1976, some bright spark, perhaps a recent Oxbridge graduate, altered the road sign to read 'Rogate et Specialiter' ('Ask and it shall be given unto you'), in time for Rogation Sunday.

Just past Rogate lies Terwick Lupin Field, which was bequeathed to the National Trust with the stipulation that it should always yield Lupins. But the Lupins were late this year, their foliage scarcely showing. Instead, I drove past what appeared to be a Wild Pear tree, a rare wonder of the woodland edge and winding hedgerow, bowed down by snow-white blossom. There was distant sunshine up on the still-grey downs.

The summit of a gentle acclivity called Cumbers Hill, past Terwick, is clothed by a copse of tall coppiced Sweet Chestnuts, with a dense Bluebell carpet underfoot. The road then runs down towards the West Sussex Rother at Trotton Bridge (as opposed to the East Sussex Rother or any other Rother). The traffic lights on this narrow bridge are always red, but that enables you to cross slowly, looking both upstream, where the water is slow and pooled, and downstream, where it ripples and races beneath Alders at bud break. Primroses, Forget-me-nots and the first Red Campions of the year were clustered on the bank beside the junction off to somewhere I have never visited called, unprepossessingly, Dumpford.

One is then obliged to speed past the drab heathers, dead

Purple Moor-grass tussocks, scattered Gorses and flushing Birches of Iping Common, as the road here is fast and straight, and out of character with the rest of the A272. It speeds you towards Midhurst, where pedestrian crossing points arrest you, and force you to notice placards advertising a Spring Fayre at The Grange, the Chichester Cathedral Festival of Flowers, and a Food & Folk Festival. Take your choice. On this day, the clock in the High Street was just three minutes slow, which by A272 standards is disappointingly accurate, for there are situations in which an element of timelessness is desirable.

Crossing the Rother again, by a weir with a spectacular plunge pool which must offer excellent Chub fishing, the road enters Easebourne, which showed off the last golden Celandine bank of the season, yet more Spanish Bluebells, and Aubretia tumbling over walls. Easebourne ends with a 'Private Bypass Bridge for motors not exceeding 2 tonnes, at owners' risk'. I used it to bypass the shadow of a dark cloud that hogged the main road – always walk in the light.

Past Cowdray Park Golf Club, and its mown verges, the road dips by an avenue of veteran Sweet Chestnuts, and up and down to a pond turned into mulligatawny soup by too many Common Carp. These bottom-feeding fish stir up sediments that would otherwise rest on the bottom, thus clouding the water and killing off photosynthesising plants. This used to be a Rudd pond, with Lilies and clear water. It needs mending, by de-carping.

Our road then leads through a wooded incised cutting in the amber sands of the Lower Greensand, and up through three ridiculously short sections of dual carriageway – sufficient to

overtake a tractor, or possibly a very slow horsebox. Vineyards
have been established on the left, a Crab Apple was blossoming, and a whole field had been surrendered to Dandelions.
All Hallows Church, at Tillington, boasts a Scots crown tower
– effectively, a cheap hollow spire on top of a tower. It was
painted, imaginatively, by JMW Turner and John Constable,
both of whom stayed at nearby Petworth House. The Park there
was laid out by Capability Brown but being a deer park it is all
but devoid of flowers, for Fallow Deer avidly consume flowers,
so much so that centuries of deer in a park lead to an acute
paucity of wild flowers. Petworth itself was brooding, beneath
giant Cedars and Holm Oaks, its narrow twisting streets hosting
an excellent ultramarine Ceonothus, a cerise Cherry in full
bloom in the churchyard of St Mary The Virgin, and the first
mauve Lilac flowers of the season.

Beyond miasmic Petworth the road straightens through pasture
fields, then rises up to Fox Hill, a wooded greensand ridge, the
summit of which provides a brief hazy vista over miles and
miles of Oak-clad countryside. All the Oaks in Sussex were
coming into leaf for me, offering the nameless yellow-greens
that only young Oak leaves know how to perfect. Oaks towered and overarched above the road as its twists past the ancient, pathless, woodland of The Mens, a place deserted.

Once, time ago, I glimpsed a yellow Brimstone butterfly wandering the spring verge there. I still remember it when passing,
which is curious as I have driven past many a Brimstone – but
that was in April 1976, in an era which made itself immortal
by imposing eternal memories. Somehow that butterfly lingers
on, as memory, within the memory of a place.

Wisborough Green is announced by the statue of a grandfather clock carved from a garden Oak struck by lightning. The clock face reads 3.45 pm: tea time, with fresh scones and medlar jelly one hopes. The village itself was entered via a riot of Dandelions along the verge. It features a classic Sussex green where cricket and stoolball are played, a duck pond with preening ducks, and a stone-faced church with a dreamy spire which beckons Swifts home. The first Horse Chestnuts were coming into bloom, against billowing dark clouds on a torn blue sky. In the dip where the River Arun flows slow beneath the road, the Oak leaflets had been touched by late frost and were discoloured by an incongruous bronze.

It was Tulip time in Billingshurst. This was celebrated by Wood Pigeons mating on a bus shelter roof and a Grey Squirrel running along a telephone line, the surest way for a Squirrel to cross a busying road. Fresh Sallow leaves, blue-grey more than green, offset the rampant mustard yellow of Wych Elm seeds on a corner before, or after, the road arch over the Portsmouth railway line – or both perhaps; it matters in precision only loosely.

Further on is an alluring turning towards the remarkably-named Itchingfield. Why did the field itch? Was it full of fleas, or biting flies, or itch-inducing plants? Any answers are lost in the sense of timelessness that the A272 instils, and ultimately matter little.

The crossroads village of Coolham boasts the only extant convent along the road, St Cuthman's – until recently there was also one in Midhurst, but that closed and is now the site of bland retirement flats where coffee mornings are held. St

Cuthman's, named after a local minor saint, has survived by becoming a retreat and meeting centre (Catholic with sacrament reserved, but all welcome).

A straight, hedged stretch leads to the only prominent rookery along the entire 30 miles, in a starkly bare Ash upon a hillock (there is, though, a rather discrete rookery in Midhurst and another at the east end of Easebourne). One is then given a choice, the sort one has to make when at a major crossroads in life: turn left for the rising woods behind Dragon's Green and the George & Dragon, or right for Shipley where Hilaire Belloc lived in a gleaming white windmill, The Countryman pub and the developing wood-pasture of Knepp Wildland, a project which seeks to rectify the division between agriculture and ecology, between time past and time present, between spring and any other time of year.

For to continue further would be to risk shattering the dream, as the A272 idyll breaks all too sudden at a busy crossroads with the A24 dual carriageway, complete with a bog-standard, standard-bog McDonalds. Continuation beyond this crossroads would necessitate entering a busy roadway that leads to the A23 arterial routeway between London and Brighton.

No, turn left, or right; or turn around and drive back towards Petersfield, and into the first sunset of May, and the month's first Nightingale night. There may even be Ghost Moths dancing the twilight verge.

Shepherd's hut in Knepp Wildland, West Sussex, 1ˢᵗ May.

In the way that we cannot be other
than ourselves even in the deep of winter
these woods where bluebells grow

are always bluebell woods.
The blunt grey tips poke up
through the melted leaf-litter.

They are entering their domain
and the light is startling to them.
They quietly jostle to get more of it.

It may be that they are concentrating
on making the blue they remember

Katharine Towers - *Bluebells*

BoPeep Copse

and I
Would arrive and go far
To where the lilies are.

Edward Thomas - *The Lofty Sky*

Below the shadow line of the South Downs, by a pond where Rudd laze away the first Lily-pad days, nestles the intriguingly-named BoPeep Copse. An ancient woodland relic, which tells of bygone times that return each spring only to go again, BoPeep Copse revels in just about all the spring plants one could expect from clay woodland, deep in the south country. Much happens beneath its crown of Pedunculate Oak and Ash, especially at spring's prime, before it slumbers the summer away in shadows numberless.

Chaffinches and a high branch Chiffchaff greeted me, a basking Grass-snake, and my first Orange-tip eggs of the year, on the sepals of Lady's Smock plants. The Bluebell's hue and perfume threatened to overwhelm and, looking up, the white fire of Wild Cherry blossom shone against a sky that only early May can perfect into such a pure azure. The last of the Lesser Celandines and Wood Anemones hugged the shady foot of a sunken bank, telling that it mattered not that spring was already starting to age.

Looking into the wood, the hulks of ancient Ash coppice stools stood out proudly, with fronds of Solomon's Seal curled above the uniform tier of dense Dog's Mercury and Herb Paris foliage. In places, Solomon's Seal and Herb Paris grew almost inter-twined; the one aspiring to be perpendicular, the other horizontal. They were coming into flower whilst the Bluebells were at peak, studded here and there with coins of Goldilocks Buttercup and Yellow Archangel. In hollows, where the Oak leaf litter was deepest, patches of soft-grey Moschatel grew, offering mid-tones to the vernal greens. Common nomad bees and spring hoverflies basked in patches of sunlight, and the first of the black and grey Ramsons Hoverflies waited, pristine, on the shining leaves of the plant they hold as theirs. They time their appearance perfectly with the flowering of their host plant, Ramsons, or Wild Garlic. Secreted amongst the bur-geoning ground-layer foliage and in the Oak leaf litter was the Garlic Snail *Oxychilus alliarius*, a whorled yellowish snail of damper places. It cleverly manufactures its own garlic, which it uses as a defence mechanism when disturbed – but that is a smelly way to identify a species. Soon the whole copse would be heady with the scent of garlic. Witchcraft would not pene-trate here, at least not while the garlic flowers.

Along the lane edge, Common Carder bees hummed their way between Herb Robert flowers, and Wood Melick grass cast strings of black pearl buds amongst its distinctive mid-green blades. Had the Early-purple Orchid been present, as opposed to being incongruously absent, BoPeep Copse would perhaps have been all too perfect for this world. All this was here and now, in a couple of acres of paradise masquerading as a hum-ble copse.

The woodland birds fell quiet, knowing perhaps that spring's flowers held sway here, with their associated insects. Yet, from a distant world outside the copse, a lone Skylark rose – *court –es-y, court-esy, curt-sy, kiss me, kiss me, kiss kiss kiss kiss –* before dissipating, into the ether, where the song-dream led.

I first walked this place as the leaf canopy was closing over, at the end of the spring of 1976. I discovered it by accident – either that, or it called me in by design. Whatever, the place had called me back again, homeward; for I had left part of my soul behind, time before, and was duty bound to return, to leave yet more of myself behind, amongst spring's flowers.

BoPeep Copse and Little Butser Hill, Hampshire, 4th May.

Suddenly above the fields you're pouring
Pure joy in a shower of bubbles,
Lacing the spring with the blue thread of summer.
You're the warmth of the sun in a song.

You're light spun to a fine filament;
Sun on a spider-thread –
That delicate.

You're the lift and balance the soul feels,
The terrible, tremulous, uncertain thrill of it –
You're all the music the heart needs,
Full of its sudden fall, silent fields.

Katrina Porteous - *Skylark*

A Foretaste of Summer

Sad, too, to think that never, never again,

Unless alone, so happy shall I walk
In the rain. When I turn away, on its fine stalk
Twilight has fined to naught, the parsley flower
Figures, suspended still and ghostly white,
The past hovering as it revisits the light.

Edward Thomas - *It Rains*

Without warning, for none was needed, a day of dull summer
weather interceded into the mid-spring riot. Air hung heavy
beneath low clouds that glimmered and glowered, offering a
calmness seldom experienced outside the deep coma of
winter, when our world is closed down. The light, or rather
lack of light, allowed a meadow to assume, prematurely, the
dark green hues of summer, with Meadow Foxtail and Timothy
grass stems standing erect, motionless, as the first summer rain-
drops pitted themselves into the dust left by a few dry warm
springtime days.

Yet, back at Avebury in deepest Wiltshire, the sounds were still
primarily of spring. Every songbird sang as those early drops
fell – and they sang both of spring and of summer, unified:
Song Thrush, silhouetted in a bare Ash, Chaffinch in a Hawthorn unfurling white buds, Blackbird up on a thatched roof
top, and a Wren entwined somewhere deep within a Box
hedge. Wood Pigeons joined in, prophesying thunder and an

apocalyptic ending to the world, and young Rooks clamoured in nests high in Horse Chestnuts and Sycamores. The rookeries had changed tune, answering and joining in with the song-dream of summer. The Swifts were beginning to reel, already. The church clock struck four, in summery somnolence.

Summer's insects were joining in too. Syrphid flies hovered above the narrow path leading into the village, beneath late-leafing Limes and above the Cow Parsley tops. They were spring species, with names like *Syrphus torvus* and *Epistrophe elegans*, but they were singing of summer. Spring's common flower bee, a frenetic insect fully deserving of the mouth-filling name of the Hairy-footed Flower Bee or *Anthophora plumipes*, ceased to dash about wantonly and assumed a more summery hum. The bumblebees joined in too, their dronings telling of lazy summer days at pollen and nectar. Only the diurnal iridescent Long-horn moths along the Hawthorn edge were silent, true only to spring.

A pair of Swallows were nesting once more, amongst the dark recess of rafters in the lychgate of St James's Church. They would raise, perhaps, two broods here; the first at Lilac and Laburnum time, now; the second within the Scabious days of August, before summer's leaving and when spring would be a forgotten dream. Here was their summer home. Swallows may have nested here for a century, as the lychgate was constructed in 1899; and longer in the church porch, for centuries perhaps. They become part of the place, integral to its being, even if they are only there for part of the year.

All manner of things seemed well within the world. The Jack-daws had even stopped stealing thatch from out of the black barn roof, and were sitting around on chimney tops in contemplative mood, perhaps as witnesses to summer's coming. Summer's shadows were here too, in incipiency. The churchyard Yew trees harvested gloom from a darkening sky, even the whitewashed walls of a thatched cottage seemed ecru-grey.

Then the rain began, pattering vertically on umbrellas that paraded their way around the standing menhirs of Avebury stone circle. It too sang of summer, being the first of the season. Its meaning was lost on a kindergarten of young Starlings feeding excitedly amongst the Daisies, on the cricket ground outfield where bad light and now rain had stopped play. The Starlings had not experienced summer rain before, or even summer.

It rained for two whole days, steadily. Throughout, the Black-birds sang, undefiled; their spirits would not be dulled. The Brown-lipped Snails took their houses on tour.

Avebury, Wiltshire, 9th May.

the summer rain is calling
gentle, constant,
almost
a whisper – like a steady breeze
through leaves, like an elongated
exhalation.

Nadia Kingsley - *Summer Rain*

The Jaws of Borrowdale

I could Invest every leaf with Awe.

ST Coleridge - *Notebook, Cumberland, 1803*

Borrowdale, up in the far Lake District, is all about rain, and the land's reaction to it – and ours. This most memorable of the Lakeland valleys is a shape-shifter, where continually-changing weather reveals myriad permutations of water, rock, slope, Bracken, tree and sky. No two visits are the same here, no two days, or even hours.

Borrowdale existed in its own rather prehistoric bubble before it was popularised by the Reverend William Gilpin and the Picturesque movement, after Gilpin's tour of Cumberland and Westmoreland in 1772. Despite a long history of mining it was only connected by road to nearby Keswick in 1842, and to the National Grid system during the 1960s.

Gilpin's hugely-successful books helped stimulate mass intellectual tourism for the young and wealthy – chinless wonders by and large, who had been set free to explore the countryside by no lesser thing than the development of the stagecoach. The Lake District quickly became an epicentre for Picturesque tourism, which determined that Nature should consist of the *beautiful* and the *sublime*. There are two sides of the Picturesque coin: sunny side up (Beauty) and tails up (Sublime). The former should be obvious; the latter helps us appreciate the darker, more frightening side of Nature – the awesome. In Borrowdale, beauty lies in the lap of horror. There is, though, a third side to every coin, however thin – the edge, which links the two sides: we ignore that, though nature doesn't. Some of us are trying to work out what the edge is – it is something to do with metaphysics, spirituality maybe, faith even.

The Romantic poets quickly followed the Picturesque into Borrowdale, for William Wordsworth was a native of Cumberland. He moved to Grasmere with his sister Dorothy in 1799, whilst Robert Southey and Samuel Taylor Coleridge came to live together at Keswick. Their house still stands, above the River Derwent, but overlooks a car park and an ugly industrial estate with its views westwards over Derwentwater, Cat Bells, and Maiden Moor desecrated. Southey loved the area, and remained there, and became part of it. Coleridge, though, quickly became a broken man: trapped within a disastrously unhappy marriage, suffering from what is now known as Seasonal Affective Disorder (SAD), and probably bipolar, he developed major problems with alcohol and laudanum addiction. But he loved the Lake District, and profoundly understood its beauty and its awesomeness. It's just that he couldn't live there, though legally obliged to. Seeking escape, he effectively invented fell walking, and even fell running. What he loved most about the Lake District was the way, 'Mist & clouds & sunshine make endless combinations, as if Heaven & Earth were forever talking to one another' (STC Notebook).

The Picturesque movement had determined precisely how beauty and the sublime should be experienced, through the narrowest and most precise of vistas. One had to stand in the right place, and look in the right direction. It was completely stage-managed, and far too narrow-minded an approach for genuine intellectuals such as Southey, let alone for profound poets like Wordsworth or deep metaphysicians like Coleridge. One of the key Picturesque features of Borrowdale was, and remains, the Lodore Falls. Here, water rages, tumbles or trickles – depending on rainfall – down a ravine on a precipitous north-facing slope before entering Derwentwater. At its fullest, it is thunderous – and petrifying. Gilpin, writing in 1772, describes it thus: 'The stream falls through a chasm between two towering perpendicular rocks. The intermediate part, broken into large fragments, forms the rough bed of the cascade. Some of these fragments stretching out in shelves, hold a depth of soil sufficient for large trees. Among these broken rocks the stream finds its way through a fall of at least an hundred feet; and in heavy rains, the water is every way suited to the grandeur of the scene.'

Robert Southey eulogises Lodore Falls in a lengthy and simply brilliant onomatopoeic poem he wrote for his children, it ends:

> Retreating and beating and meeting and sheeting,
> Delaying and straying and playing and spraying,
> Advancing and prancing and glancing and dancing,
> Recoiling, turmoiling and toiling and boiling,
> And gleaming and streaming and steaming and beaming,
> And rushing and flushing and brushing and gushing,
> And flapping and rapping and clapping and slapping,
> And curling and whirling and purling and twirling,
> And thumping and plumping and bumping and jumping,
> And dashing and flashing and splashing and clashing;
> And so never ending, but always descending,
> Sounds and motions for ever and ever are blending,
> All at once and all o'er, with a mighty uproar;
> And this way the water comes down at Lodore.

To the west of the Falls stands an unscalable crag, from out of which hockey stick Ash, Oaks and Sycamores grow. They managed to establish themselves in crevices in the vertical face, away from browsing animals, and grow out horizontally before bending and starting their aspirational ascent.

To the east climbs an Oak-clad slope, barely ascendable, carpeted in ferns, boulder mosses, clumps of Bilberry, Great Wood-rush and Wood Sorrel, and pocketed by vacant hollows of last year's Oak leaf litter. Something approximating to a path scrambles up this slope, though in heavy rain it becomes a tributary of the Falls. Halfway up I realised I could not return that way, it was too steep; I carried on regardless, upwards and onwards. Coleridge would have done likewise, but without stopping to consider the predicament.

My visit occurred at the back end of Borrowdale's equivalent of a prolonged drought – a dryish fortnight, and in mid-May to boot. Wood Ants scurried across the ground, unhindered for once by the rivulets or cascades that normally impede their movements. The Falls were running short of water, and were gasping, rather than roaring. One could hear oneself think, and even hear the trilling of a Wood Warbler somewhere in the Oak canopy.

Here and there were casualties of war – Oaks and Larches felled by autumn's storms, root plates upended. They were left to rot, for here no chainsaws roar – no one in their right mind would carry a chainsaw up this slope (yet someone had been daft enough to plant Larch trees up here, in full knowledge that the timber would not be harvestable – a tax dodge probably). One Larch trunk had managed to prostrate itself along the only straight stretch of ascending path, necessitating a 20-metre plank walk, with a crevasse beckoning westwards. Higher up, I was tempted by a short cut, along another prostrate trunk – above a 15-metre sheer drop, but no longer being 12 years old I decided to forgo that particular challenge and followed the path around a tortuous corner, allowing my spirit to take the short cut instead.

Rain began suddenly, borne on a windless stillness, and drowning the low sound of the Falls. Clouds ghosted in, descending, wreathing the mountaintops before enveloping the sides – Cat Bells and Skiddaw, across and above the leaf-shrouded lake, vanished, as if they had never been. The vertical horizon dropped by 100 metres in a couple of minutes. Then the sound of running water began, making its way down the mountain; gathering itself in volume and pace, awakening the waterfalls of Lodore. Within minutes the parched boulder mosses had once again become sponges, giant slugs were crawling everywhere, and I had little idea where I was. The only thing to do was to join the water, and descend – somehow, anyhow. Had I continued upwards I would probably have entered Narnia.

Coleridge described Lodore in his notebook: 'The precipitation of fallen angels, flight, confusion and distraction – all harmonised into one.' He understood the Sublime – way beyond the superficial understanding proffered by the Picturesque movement. If anything he delved too deep, and disturbed his own mind.

And with the rain came an early darkness, in which the Yew trees around the Lodore Falls Hotel once again became obsidian sentinels of winter, and a lone Copper Beech ached of November. Steadily, as a Blackbird warbled softly, almost apologetically, the coral-pink petals were washed off the hotel garden's Cherry trees, to wend their way into the depths of Derwentwater beyond, and drown.

In the valley bottom, around the eroded spot from where the Falls are supposed to be viewed, giant boulders lie beneath mighty girth Oaks contained by Ivy growths as thick as a man's thigh. And on one boulder, placed assiduously – a half-drunk cup of Costa coffee, left there as a warning to nature, of man's intent towards it.

I had experienced the sublime, and it is real. But like Coleridge I needed to venture deeper. We all do.

Lodore Falls Hotel, Borrowdale, Cumbria, 17th May.

Wildland Dawn

How is it, Shadows! that I knew ye not?

John Keats - *Ode on Indolence*

It began with an awakening, a rude and yet joyful awakening, at the imminence of dawn. Somewhere in the distance, from the village of Dial Post, a cockerel crowed – once, twice, and then, thrice. Then a Cuckoo called, dutifully, from a faraway shadowland of Oaks in the monastral west. Out there, shapes and patterns were beginning to form, from nothingness, from nebulous opacity, from shadows intensely numberless, heralded by a ghostly light which seemed to come from the ground, for frost had formed over dew-laden grass and was giving rise to a curious distortion of light – call it lightfulness – even as the frost crystals dissolved back into water. Layers of mist vapours added to the pallid light, adding tiers of colour-lessness, which changed continuously, until colours came into being. All this was seen through the smallest of windows, for such is our experience of the world.

Inside a shepherd's hut at Knepp Wildland the air was perishing cold, death-like almost, yet it was moving; as if it was being taken over by some unstoppable force. Something was trying to break in from another dimension. The stone in front of the tomb was being rolled away, presumably by an unseen angel who had stolen in from the great elsewhere. Within seconds the walls of the hut were vibrating, shaking almost, under the mighty onslaught of what we naively and blandly call the dawn chorus. It was total bombardment, a wall of sound and airwave movement in which it was impossible to separate individual songsters. Here at last was the full unity of being.

'Dawn chorus' is a pathetically inadequate term for this unified eruption of ecstatic joy, for a mass hymn of thanksgiving spontaneously offered by, or at least on behalf of, all life forms. The birds were the choristers, yes, but they were by no means alone. It felt almost as if the earth was heaving, or at least that it wanted to. Sounds were being made that are way beyond the limitations of our hearing, and were combining, through interfusion, to form something beyond our comprehension. We need antennae, to start to understand, for so much of what we categorise as being sound is actually airwave vibration, which we perceive only subconsciously.

The birds' message was, simplified right down into a simple poetic language: "Once, we suffered winter, now we dance the spring!"

There was only one thing to do, join in. Here though, language – even the ecstatic English language – fails miserably. We cannot express that level of joy, that purity of ecstasy, that wondrousness of being, that transformation from being into better being. It is not simply that words fail us, for religion fails us here too: faith itself breaks down, degenerating into an inadequate framework within which the song-dream of each spring dawn can be scarcely experienced. Perhaps Nature, which goes beyond right and wrong, perfection and imperfection, and just *is*, provides a better framework?

The pupa case was breaking open. There was nothing, nothing left of me. I had grown wings, the wings of today. I flew.

On returning, from wherever I had been transported to, I scrambled for the solitary book I had brought with me, Edward Thomas's *Light and Twilight*, and read his unsurmountable description of the dawn chorus: 'I knew that multitudinous song, emparadised, remote, serene, only it had never before been so fair, nor had thus changed all the desires and joys of the flesh into this aerial sweetness on keys of ivory and crystal, pipes of reed and gold, strings of gossamer, wreathed horns of exquisite shell, warm throats of beauty, of love, of youth, of joy. It could not know change. It would build fair towers, soaring high into the dawn and into evening, and from their windows would beckon all the lovers of the world, and all the minstrels, glad and at ease in eternity.'

Shepherd's hut at Knepp Wildland, West Sussex, 23rd May.

The tawny owl wakes us to our widowhood.
The dawn is the chorus.

The dawn is completely composed.
The pens of its beaks are dry.

Day will never sound the same,
nor night know which song wakes her.

David Morley - *Chorus*

Nether Stowey Revisited

And 'mid these dancing rocks at once and ever
It flung up momently the sacred river.

ST Coleridge - *Kubla Khan*

The deeply-atmospheric Quantock Hills in West Somerset are formed from a band of ancient red Devonian Sandstone that reaches north-westwards from Taunton towards the Bristol Channel and far-flung Wales. Several hills within the range exceed 320 metres in height. Many of them are separated by steeply incised combes and stream valleys. All are clad in expanses of Heathers, Gorses and Bracken, with stunted Oak woodland dominating the lower slopes. Most of the hill range is open common land, haunted by plaintive Meadow Pipits and attendant Cuckoos. One large part has been brutalised by alien conifers, and sulks, a place apart.

The Quants, as they are known locally, are a relatively small Area of Outstanding Natural Beauty and a moderately large Site of Special Scientific Interest. These designations belittle their main significance, for they are the birthplace of the globally renowned Romantic movement, and as such must merit designation as a World Heritage Site, but this most special place is overshadowed in terms of literary fame by the Lake District. To most people, the Quantocks are dwarfed by their more famous neighbour, Exmoor, of which they are effectively a satellite.

It all started here, on New Year's Day in 1797, when the young poet, metaphysician, Unitarian preacher, and radical thinker Samuel Taylor Coleridge moved with his wife, and baby son, to a primitive cottage in Nether Stowey – a backwater village mid-way along the Quantocks' eastern flank. Originally, he had planned to rent a millstream cottage a little to the south-east at Adscombe, but had to take the three-up three-down hovel in Stowey when the mill lease fell through. Coleridge (who hated being called Samuel, loathed the name Taylor, detested the combination of the two, and preferred to be called plain Coleridge or STC) was attracted to Stowey by a fellow free-thinker, Tom Poole, who owned a tannery business. They had met at lectures the 24-year-old Coleridge had delivered, in Bristol – then a hotbed of progressive and anti-slavery think-ing. Coleridge was seeking some sort of self-sufficient hippy-like existence, where he could study, write, think, commune with nature, and above all – dream. He emanated from Ottery St Mary, in Devon, but had spent much of his life as an all-year-round boarder at Christ's Hospital school in central London. At Stowey he made a haphazard and naïve go at self-sufficiency, keeping ducks, geese, and pigs, and cultivating corn and vegetables. Poole, who lived close by, provided much moral, intellectual and practical support.

Meanwhile, some thirty miles to the south, 27-year-old William Wordsworth was living with his younger sister, Dorothy, at Racedown Lodge, to the south of Crewkerne, earning a living by bringing up a wealthy widower's child whilst struggling to break through as a poet. Coleridge and Wordsworth first met in Bristol in September 1795. Then, in early June 1797, after preaching in Bridgwater the ebullient Coleridge hastened to Racedown, probably on the spur of the moment. The trio hit it off so strongly that Coleridge stayed there for two weeks. Then, ably assisted by the wealthy and well-connected Poole, the Wordsworths were installed in a rented villa at Alfoxton near Holford, about four miles to the north of Nether Stowey, along the Quantocks' eastern edge. Dorothy maintained a journal of her time there.

During that summer and autumn, Coleridge and the two Wordsworths rampaged all over the Quantocks, by daylight and moonlight, as experiential and fiercely intellectual children of nature. Few people can have loved and used a summer so well, living within its moments and allowing its moods to direct their movements. They were particularly interested in the relationships and synchronicities between nature and the human mind. Almost childlike, they delighted in tracing streams seawards; but all the while they were seeking nature's meaning, and following summer's course. The striated shore platforms and banded cliffs of shales and limestones around Kilve Beach fascinated them. Kilve has been a place of pilgrimage for poets ever since. Coleridge was the most adventurous, roaming alone as far as Porlock, where *Kubla Khan* was composed (under the influence of two grains of opium).

Other like-minded radicals visited them, staying in Coleridge's cramped cottage – notably Charles Lamb, William Hazlitt, Robert Southey and the Jacobin sympathiser John Thelwall. The latter was a marked man politically. His visit to Stowey was ill-advised, especially as it was known that Wordsworth had been to France during the Revolution and had expressed support for its original ideals. The young poets were watched by a government spy and treated with severe suspicion by the locals.

Nonetheless, Coleridge composed much of his best verse whilst living at Stowey: *The Rime of the Ancient Mariner, Kubla Khan, This Lime-Tree Bower My Prison,* and *Frost at Midnight.* Most significantly, the first edition of *Lyrical Ballads,* the joint venture between Wordsworth and Coleridge which transformed the world of poetry by shocking it out of a post-Milton and Picturesque-fantasy hangover, was produced during this era. The two poets sought to write as one – Coleridge had the greater imagination, Wordsworth was the greater wordsmith. The following astounding lines from Wordsworth's 'Tintern Abbey', first published in *Lyrical Ballads,* represent the glorious fusion of Coleridgean vision and Wordsworthian poetics (some lines are pure Coleridge, others are classic Wordsworth), and constitute a definition of nature that remains definitive –

> For I have learned
> To look on nature ... And I have felt
> A presence that disturbs me with the joy
> Of elevated thoughts; a sense sublime
> Of something far more deeply interfused,
> Whose dwelling is the light of setting suns,
> And the round ocean, and the living air,
> And the blue sky, and in the mind of man,
> A motion and a spirit, that impels
> All thinking things, all objects of all thought,
> And rolls through all things. Therefore am I still
> A lover of the meadows and the woods...

But the dream existence did not last, as Wordsworth found himself unable to renew the tenancy at Alfoxden or find alternative accommodation elsewhere in the district – undoubtedly due to social and political concerns. So, in September 1798,

Wordsworth, Dorothy, and Coleridge travelled to Germany for a philosophy study trip that was scheduled to last three months but extended itself to ten. Coleridge's second son was born, and died, during this time. He never saw the infant. His wife, Sara, never forgave him. On returning from Germany the Wordsworths, and then the Coleridges, relocated to the Lake District: to Grasmere and Keswick, respectively. However, Coleridge never settled there and steadily fell apart mentally. His marriage degenerated into disaster, for this was long before divorce was feasible. Steadily, even his bond with Wordsworth broke.

Today, some of the magic the poets loved so deeply, and responded to so profoundly, lingers on in the Quantocks, at least for those who seek it. Edward Thomas, a great admirer of Coleridge's mind, sought it out. His pilgrimage, recorded in his rural prose-poem *In Pursuit of Spring*, his greatest rural prose-poem, took in Nether Stowey and Kilve Beach, before ending on Cothelstone Hill in the southern Quantocks on Easter Day in 1913. There he placed a bunch of spring flowers on a spot he felt represented winter's grave: 'I had found Winter's grave; I had found Spring…'. But apart from the National Trust owning what is now known as Coleridge Cottage, an excellent Coleridge Way long-distance footpath, and Alfoxton remaining remarkably unchanged, there is scarcely any vestige of Romantic imprint on the place; one particularly tranquil spot is known as Dorothy's Waterfall, though it is shown on few maps. Instead, the Quantocks have become dog-walking country.

Conifer slum apart, the Quantock landscape itself has remained relatively intact. It is a place of quietude, crossed by only two meandering lanes – though bordered below by cumbersome main roads along which heavy traffic throbs. The Sessile Oak woods stand long neglected, though they were vigorously coppiced when the Romantics knew them, and supplied bark to Poole's tannery. The trunks, multi-stemmed from past coppicing, now grow twisted and stunted, and hang heavy with epiphytic lichens, mosses, and ferns. Once the canopy closes over in late May these woods become increasingly dank, dark, restraining, and fungal, whereas 200 years ago they would have abounded with wild flowers and butterflies, and pulsated with bird song. The open commons now support only a few hill sheep, which congregate in certain areas, in hefts, such that parts are deemed by nature conservationists to be 'too heavily grazed', others 'under-grazed', whatever those terms mean. Two hundred years ago the commons were heavily grazed, by cattle and sheep, and extensively burnt, and consequently far more open. Vistas have closed in, woods have marched uphill and coconut-scented Gorse has burgeoned, whereas for centuries European Gorse foliage was cut for winter fodder and its stems for fuel. Bracken was also harvested for winter bedding. Both Bracken and European Gorse would also have been checked by severe winter weather and by late frosts – *Frost at Midnight* weather. Recently, swathes of the northern Quantocks have been invaded by *Rhododendron Ponticum*, which was first introduced to Britain in 1763 as a garden shrub, necessitating large scale clearance for conservation reasons.

And, as elsewhere, the flowers and grasses have largely gone from the meadows around Nether Stowey, replaced by heavily-fertilised Rye grass monocultures of uniform acrylic green. The two poets, and Dorothy, were competent naturalists (though without today's books or knowledge), and would be horrified and mystified by the absence of these flowers today; most notably the disappearance of the formerly ubiquitous Buttercup. Today's monocultures of emerald green Rye grass leys, devoid of flowers, would appal them. They should half-recognise the North Devon cattle that graze some of the fields by Nether Stowey, but would be amazed by the size of today's Holstein and Charolais beasts. They would be acutely alarmed by the paucity of birds and bird song: in the woods, on the commons, and especially in the fields. Gone too are the people who dwelt upon the land, who fascinated Wordsworth – the commoners, farm labourers, woodmen and itinerants; now replaced by mountain bikers and dog walkers, as the Quantocks has shifted into being a recreational landscape. The conifer forest would probably be abhorrent to the Wordsworths, but would attract Coleridge, who would recognise it for what it is – a hell on earth, representing the darkest shadows of the mind.

But above all, above the distant glowering of Hinkley Point Nuclear Power Station, and the eructation of smoke from the Port Talbot Steelworks across the Bristol Channel, what would alarm them most is today's traffic noise – which they would have to climb far into the hills to escape. It would kill their musings, and shatter their experience of ethereal moments of being – even as it does my own, and in all likelihood, yours.

They would have to narrow their vistas and dream-lines right down, in order to be able to concentrate – and confine their ruminations to the quieter, remoter places. Without doubt, they would find themselves closest to nature in the depths of night, though a multitude of stars have faded from our light-polluted night skies.

This is what has befallen nature lovers in recent decades: we are living in a world which has fallen still further from grace, for nature's church is now being rapidly eroded away. If a stand can be made anywhere, it is in the Quantocks, but the place needs first to recognise its significance, re-brand itself, and recognise its true identity.

The truth is that until relatively recently, in the time before modern comforts, people lived for spring and summer – with an amaranthine yearning which helped them survive. Coleridge Cottage was cold and damp, and on at least one occasion it flooded spectacularly. Clothing and footwear were far less adequate than what we are used to today. Winter was literally hellish. But that made spring and summer all the more glorious.

Here, in the Quantocks, spring and summer coexist for a precious while, as ever. Summer comes early to the warm valley bottoms and to Nether Stowey's pleasant gardens; spring's flowers fade early there and are replaced during late May by summer's. Whilst high up on the hills, above the 300-metre contour (the old-fashioned 1,000 feet mark), spring comes late and lingers long – windswept Hawthorn bushes come into flower in mid-June, long after their valley bottom counterparts have cast their blossom and started to conjure their autumn colours, the Bracken croziers unfurl on the high hill tops weeks after their valley bottom cousins.

Somewhere, up there in the Quantocks, now lies Spring's grave, upon which summer's flowers have been cast. It would take a poet to spot it.

Nether Stowey, Somerset, 30th May.

Hawthorn Days

At hawthorn-time in Wiltshire travelling
In search of something chance would never bring

Edward Thomas - *Lob*

Early morning clouds were fast burning off, enabling the early June sunshine to dazzle from the chalk surface of the sunken track that leads up to the heights of Cherhill Downs, on the Marlborough Downs. On both sides white Hawthorn blossom dazzled further, almost blindingly. It was not as white as snow; rather, snow attempts to be as pure as Hawthorn blossom, but fails – not least because at Hawthorn time the thought of snow pales into insignificance. Yet the Hawthorn combined with its neighbouring Cow Parsley to give off a curious scent of putrescence – for winter's slain and forgotten corpse was rotting, even as spring's petals now littered the ground; and out on the vast surrounding acreages of ploughland, the barley heads were silvering.

It all seemed too much for the morning Skylark, who ascended once more into the ether, in apparent or actual ecstasy. The day's first butterfly was a Brimstone: a female born the previous August, over-wintered in a Bramble or Ivy entanglement, mated during early rays of warm spring sun, and now wandering ceaselessly in search of Purging Buckthorn bushes on which to lay the last of her pale elliptical eggs. She would seek out bushes in dynamic growth phase, such as those regrowing vigorously after being mindlessly mown down by some wanton machine. Her days were numbered.

A Red Admiral hurtled by, heading due north with the full purpose of an immigrant butterfly. The dream was becoming real, as all true dreams should, and this butterfly would not be denied. Others would follow. It was to be a Red Admiral summer.

Up on the downland summit the bridle gate was singing its rattling song in a rising south-easterly breeze. The rusting metal label stated that it had been made by AJ Charlton & Son Ltd, of Buckland Down, Frome, in the county of Somerset. Had they intentionally made a singing gate? Probably not: it is likely that the hilltop had modified it, bent it to its purpose. Such is the power of spirit of place.

The coarse grasses had grown rampantly in the spring rains – Tor-grass and Upright Brome grass – and had covered over the Early Thyme now struggling to flower on the ant hills. Then, alarmingly, a Meadow Pipit took off from below my feet; and there, in the secret space between two outsized clumps of Cocks-foot grass, nestled four russet brown eggs in a neat nest of twisted dead grass blades. Meadow Pipit nests are almost impossible to find intentionally, except when young are being fed – to find the eggs, one has almost to tread on a sitting hen, inadvertently. The feeling of guilt, in such situations, is perhaps unsurmountable, for such experiences remind us that we are despoilers of Creation: one photograph, for memory, and quickly out and onward. Then another Lark rose, in song, and the guilt dissipated, along with the last of the morning's clouds.

Somewhere, elsewhere, in the distance, Sunday morning's

church bells were sounding, gathering the faithful in, for the redemption of sin and the purification of the human soul by church servicing. But up here, on Cherhill, one of the faithful was already worshipping – following the song-dream. The morning of the month offered such oneness.

Silver the Brome heads; muted purple the Cock's-foot; haughty the Tor-grass; and yellow the pollen, prolific on the anthers of the Glaucous Sedge carpet. The hay fever season was opening, as the grass heads wavered constantly in an anticyclonic breeze, casting their pollen wilfully. Pulses of pollen took off from along the windward slope with every penetrating gust.

Hidden in the head of a deep combe lies a copse of Beech and Sycamore. There, cattle browse, cud, and shade away the summer heat. Leaf-cutter bees were cutting out rolls from the fresh Beech leaves, with which to line their nest holes, whilst the males of the common dagger or assassin fly *Empis tessellata* hunted out smaller insects to offer as gifts to their intended mates, and early summer Syrphid flies feasted on nectar offered by dangling Sycamore flowers. Summer's offerings were being exploited here.

In the Coombes of Calstone, the hinterland combes, fringed with Buttercups yellow and Pignut white, the year was running ahead. On the steeper, sunnier slopes, where the chalk soil thinned, June's Orchids were coming into bloom – Common Spotteds and the earliest of the Fragrant Orchids. Soon these slopes would turn pink with Orchids, Thymes and Clovers; but now they were yellow, with Common Birds'-foot-trefoil, Common Rockrose and Horseshoe Vetch. Where the soil was deeper, the Lesser Butterfly-orchids offered off-white flowers (with diagnostic parallel pollinia) and blue-green leaves that almost hinted of faraway summer seas.

And everywhere there were insects, the insects of early summer. The first of the Burnet Moths were emerging – the Narrow-bordered Five-spot Burnet, from prominent straw-like cocoons constructed high on the sturdy remnant stems of last summer's Knapweeds. Fat, black bodied, with scarlet and black wings, they flopped on to flower heads to feed, before commencing their incessant whirring explorations amongst the nodding grass heads. A mating pair nestled, end to end, motionless, amongst the grass litter layer.

And sudden, shimmering psychedelic blues and greens amongst the waving grass heads, the insect that forces us to believe in fairies – the iridescent-winged Cistus Forester moth, a day-flying beauty that seems, and is, too lovely to be of this world. They were there, suddenly, everywhere, weaving their magic trails, never rising above the grass-head canopy, searching endlessly. Below them, on the banks, blundered the olive-green and brown-winged Garden Chafer, whose larval grubs feed greedily on grass roots; and here and there, ever-plodding purposefully along, the Bloody Nosed Beetle, a blue-black minibeast that grazes innocently on Bedstraw foliage but protrudes a red warning fluid when wilfully disturbed; and everywhere, worker ants a-busying themselves.

There and here were butterflies aplenty: Marsh Fritillaries, with stained-glass-window wings, disputing possession of the Buttercup flowers and spiralling upwards together, contesting possession of the air itself. Too long they had been caterpillars, now they claimed the living air. They abounded along the bottoms of every combe; the males moving around in loose groups, seeking out virgin females hiding in the grasses, and squabbling amongst themselves. Their name is perhaps a misnomer, for they are as at home here on the western chalk as they are in any marsh or mire.

Spring's songbirds were surrendering the song-dream to summer's insects. Down in the delving combe, where the heat steadily intensified, summer was being born. The first grasshopper of the year, a Meadow Grasshopper, began to call, faintly and almost apologetically, as if from the future. Slowly, the grasslands began to hum, for everywhere insects were busying themselves – ants, bees, flies, beetles and bugs, butterflies and day-flying moths. The song-dream was loud and clear here, enhanced by the prayer-wheel call of a nearby Corn Bunting and the distant jangle of a cock Yellowhammer, somewhere up in the Hawthorn snow. Calstone Coombes were turning into paradise, as summer approached, dressed in weeds and borne on iridescent wings.

Some two hundred years ago Samuel Coleridge was living in the sleepy town of Calne, two miles westwards, though the town now scarcely remembers him. He was recovering from a marriage turned sour and conquering addictions to alcohol and laudanum; and dealing with a brilliant mind scarcely able to determine fantasy from reality or concentrate on anything for long. Calne was kind to him, such that he brought together his best collection of poetry whilst living there, his *Sibylline Leaves*. A prodigious walker, by day and night, he undoubtedly wandered up to Cherhill Downs, having travelled past many times on the stagecoach road that still skirts its northern edge; and being one who seldom twice ventured the same way, he would have returned through the deepening combes of Calstone. At some point during 1814 or 1815 he left the following thought in the notebook that covered those far-away years: 'If a man could pass thro' Paradise in a Dream, & have a flower presented to him as a pledge that his Soul had really been there, & (if he) found that flower in his hand when he awake – Aye! And what then?'

Yes, Coleridge had wandered up and over Cherhill and back down through the mazy combes of Calstone; and in early summer. That musing on the metaphysics of paradise and existence had to have hit him there, and nowhere else, because it feels imprinted in the collective memory of the place. Of course, there is no written record of the pilgrim Coleridge ever venturing there, yet Cherhill's slopes and Calstone's combes appear to retain some memory of him, and those who know how to look will find some part of him ever there, especially when summer is a-coming in. But you must look with your inner eye; look with a poet's eye – and in doing so make contact with those of similar disposition who have left part of themselves behind; and on departing, you must leave some part of your own soul behind, as a gift to a place which collects human memories and adds them to its own. For places suck us in, inwardly knead associations, then spit us out, but retain the memories as theirs, and share them as fragments cast to passers-by.

Cherhill Downs and Calstone Coombes, Calne, Wiltshire, 4th June.

Where lemon melts to mustard
its song a throaty clamour,
to every month a new moon,
each hedge, a yellowhammer.

Alison Brackenbury - *Territories*

Battle Wood

I found the poems in the fields
And only wrote them down

John Clare - *Sighing for Retirement*

Somewhere in the deeply-folded and near-timeless landscape
to the south and east of Hereford – beyond the Golden Triangle
Daffodil lands – there is a wooded hillock on the Woolhope
Limestone, called Battle Wood. It is surrounded by an undulating terrain of unfrequented woods and randomly thrown
sheep fields. It is a secret place, so secret that its name has
been sympathetically changed here, to afford it the anonymity
it openly craves.

It was early June: spring had faltered and summer was in tardy
mode, leaving the natural world in a time vacuum which was
being filled by pulses of rain that spoke of no season in particular. Bird song greeted me as I entered the wood at the rain's
end – Blackcap, Robin and fading Willow Warbler, all hidden
amongst Sallow thickets and impenetrable Brambles. Dampness was rising from the woodland ride, evaporating even
while the soft grasses hung heavy. A Speckled Yellow moth,
disturbed by my presence, blundered off aimless into mid-
morning sun, losing itself somewhere there. The day's first flies
took to the air, as the morning began to warm. A Wood Pigeon
clattered off a nest near the top of a young Silver Birch. A distant ewe called for an errant lamb, reminding me that there
was a world outside this wooded place.

This was a time of mending, as the night's deluge faded beyond memory. Here indeed is a place of mending, for the deep wounds of 20th-century forestry are being healed on this small hill: the alien conifers that frowned on the woodland crest have been cleared away, and the wood has set about regenerating itself, with saplings of Ash, Silver Birch, Sallows, Pedunculate Oak and Wild Service Tree, where the dismal conifers had placed winter into spring and summer.

Yes, that's it: the principal objection to non-native conifers is that they impose a perpetual winter on the places they have infested, like the White Witch of Narnia. They slaughter spring and maim summer, even in somewhere as sacrosanct as ancient woodland. But the light shines in the darkness, and the darkness will never overcome it. Now, the Bluebells, Common Dog Violets and Yellow Archangel are reasserting themselves in Battle Wood, which has come under new friendly ownership; and along rides where pheasant feeders once stood, haphazard and gloomily, nest boxes and Dormouse boxes welcome nature back.

Halfway along the straight and narrow ride that contours the wood's south-facing slope lies an open area of calcareous grassland, offering Dropwort, gentian-blue Common Milkwort, spent Cowslips, Quaking-grass and vibrant yellow Common Bird's-foot-trefoil and Yellow Meadow Vetchling underfoot, all fringed with the yellow-greens of Bracken fronds, Wood Spurge and Brome Grass, which reflect and glorify the colour of leafing Oaks.

Then, from out of a stand of Oxeye Daisies, budding and bursting, offering purity beyond whiteness, the butterflies took off.

Theirs was the morning skies and the woodland rides and clearings, and theirs forever is the deep blue of the Bugle, their esteemed favoured flower. They were Pearl-bordered Fritillaries, brassy denizens of woodland clearings rich in Violet haze; their role was to retake what had been destroyed, and rekindle the spirit of the place. Their graceful flight, low, swift, dance-like and ceaseless over the woodland clearings, was announced by a rapturous Song Thrush and blessed by the wayside priest of the springtime trees, the cock Blackbird. They had come home.

Here, the restorative power of nature, freed from the tyranny of materially-minded man-unkind, was reasserting itself, and all was coming well with the world; even as the sun climbed and promised summer, and the purple heads of Wild Columbine nodded wisely over the Bracken.

So much depends on – not so much our ownership, but – our stewardship of land. Few of us understand this, let alone accept it.

Then, along the Hawthorn lane, in deep warmth, the year's first Bramble flowers opened. Summer was being born, after the siege of Battle Wood. The song-dream called from every flower, leaf, spray, hillock and cloud citadel in the sky.

Woolhope, Herefordshire, 6th June.

I could become tree and twig,
songbird and owl, and learn to know nothing of what feet feel
from the ground, if I lay down in the rain, now, under the trees.

Angela France - *Now, Under the Trees*

On Selsley Common

the winds turn soft
blowing white butterflies
out of the dog-rose hedges
Laurie Lee - *The Three Winds*

It was not yet mid-June, and distant-speck Skylarks were still proclaiming the coming of summer. Yet the Beech trees lining the foot of the west-facing escarpment of the Cotswold edge held boughs heavy with fully-swollen nuts and dark blueish leaves that told ominously of September. The trees held these offerings almost apologetically. Their roots spread deep into advancing time.

Beginnings and endings are often intertwined; and this was the beginning of summer's ending, even as midsummer's Fragrant Orchids were coming into bloom amongst a mustard haze of Yellow-rattle turf, the Ribwort Plantain heads were fringed with white anthers, and the first male Meadow Browns were bib-bobbing over still-verdant grasses, where dusky Small Blue butterflies basked head down, low down. A hatch of those remarkably cryptic bumblebee-mimic hoverflies *Volucella bombylans* had taken place – two-winged versions of the familiar Red- and White-tailed bumblebees. Summer's Meadow Grasshoppers were starting up. They were a week late.

At that moment of shocking realisation, spring became a forgotten dream, and autumn an inevitability. There is so much we have to deny, so many things we must turn a blind eye to in pursuit of happiness, so much that can sadden us in nature.

That may be one reason why our relationship with nature has dwindled so much – there is too much pain intertwined with the joy, too much sadness threatening to shatter the ecstasy. It's the sublime in nature that hurts us, that frightening juxtaposition of beauty and awe. We seek a joy without pain.

Nature is amoral, and doesn't have feelings of its own and doesn't care about ours. It just *is*, existing obliviously amongst its myriad unconnected moments of being, a dynamism of change. As individuals, our relationship with nature is a marriage, and in successful marriages there is much we have to turn a blind eye to, and much we must forget: indeed, marriage rather explores the art of growing apart together.

Then, in the middle of the sunken footpath that leads diagonally upslope, rested a brace of succulent-looking Roman Snails that recent rain had brought out of their usual summery torpor. A large and garrulous party of ramblers was heading that way, at speed. At this point I chose to break the law, flagrantly: I picked the snails up and moved them out of harm's way – it was either that or a lengthy explanation, repeated several times. Technically, a license was required for that action, as the Roman Snail is a protected species and handling them must be licensed, but in this instance there was neither the time nor, it must be confessed, the inclination, for this was neither the time nor the place for bureaucracy. I acted in the spirit rather than the letter of the law, believing that 'The letter killeth but the spirit giveth life.' Follow the spirit, follow the song-dream.

It was time for the first ice cream of the summer.

Selsley Common, Stroud, Gloucestershire, 9th June.

Manifold

The poetry of earth is never dead

John Keats - *On the Grasshopper and Cricket*

Deep in White Peak country, the Manifold River had dried up,
or rather it had sunk underground at a swallow hole in the
Carboniferous Limestone called Red Hurst, and was doubtless
meandering through caverns measureless to man down to a sun-
less sea – or something like that: anything to escape from the
deus ex machina of today's Anthropocene world. The forsaken
riverbed consisted aridly of boulders, pebbles and a curious
grey powdery dust, all overhung by dusky Sycamore boughs.
A pair of Grey Wagtail picked their way through, finding little.
The main river life had gone upstream or downstream.

A stillness rarely encountered in summer enshrouded the valley, and was almost overwhelming. It was complemented by a
quietude broken only by brief outbursts of bird song and the
far distant eructation of maternal sheep. Up here spring and
summer happen later than down in the south, or indeed downhill or rather, down dale: the Hawthorn was just coming into
full bloom, pink or white, or both; the russet heads of Meadow
Foxtail grasses cast pollen whenever the stillness was disturbed,
in marshy hollows small dark hoverflies gathered on the pink
flower heads of Common Bistort, and Water Avens cast their
pale purple heads in sweet surrender to summer.

The Wych Elm leaves told me I was in the North, for there they are shaped and postured more angularly than in the South. And here, many Ash trees had only recently sprung into leaf, heralding the start of summer in the Dales. It was mid-June.

I had expected to hear the laughter of a running river, singing of summer with full-throated ease, but heard only the roaring silence of summer coming. Then a tractor arrived to mow two acres of riverside meadow grass that wanted only to be left in peace, unmown, and the dream was fully shattered. The driver told me he wasn't going to bother taking the hay away – just tidying the valley up. There are times one has to escape from human interference.

Crossing over the plateau into Dovedale, by way of the village of Wetton which was being terrorised by gigantean tractors full of self-important contractors fuelled by Red Bull, I found the River Dove in full sweet song. I had been in the wrong valley. Such mistakes are easily made, for it is hard at times to deter-mine which place is calling you – the lines get crossed. Here, Mayflies were casting themselves onto the water, as their ephemeral lives were ending, it being early evening. Fly fish-ermen were out, at peace with time and place. I watched a trio of spent Mayflies fall wantonly onto the water, then followed them as they spluttered downstream, ignored by any rising Brown Trout, to lose themselves in the soulful Dove. It seemed a sensible thing to do: to drift into eternity, and sink. Along the wood edge, a cock Redstart was feeding Mayflies to young, in a nest secreted away in a hollow in an ancient Sycamore.

A Dipper chanced by, a bird I seldom see or hear – and so app-reciate them the more when they do appear. I had determined to encounter this underwater-swimming bird here, which bibs and bobs on mossy river boulders before skimming off at speed. It sped by: five seconds worth, before hazing into memory. If such experiences are not noted down they are all too often forgotten. Always keep a nature diary: it will be a comfort to you in time to come.

The river here is in the slow process of cutting itself a limestone gorge, but rivers have time on their side. Wordsworthian 'lofty crags' towered sunwards, mostly wooded, but with stark crags and pinnacles here and there. Eastwards stood a slanting slope of Fescue grassland and scree, too steep for ageing knees. It was studded with the mounds of Yellow Meadow Ants, already turning brown in summer heat, and yellowed with the flowers of Common Rockrose, Common Bird's-foot-trefoil, Crosswort, and more Buttercups than eyes can count. It was remarkably lush and grassy for steep slope grassland on limestone, telling of a climate inclined towards dampness.

Butterflies were going to roost in sunlit grassy hollows along the lower slopes: Dingy Skippers on Salad Burnet and Ribwort Plantain heads, wrapping their wings around stilled flower heads; Common Blues loosely aggregating, face down, on Cock's-foot heads; and Small Heaths randomly and singly in short grass pockets. Each species had its own roosting strategy.

The river was lulling the valley, and itself, to sleep, as Jackdaws circled high overhead, in silence and somnolence. A great peace was promising to descend.

Sitting, set back against a pollard Ash, surrounded by blue-eyed Germander Speedwell flowers that were closing up for the day, I determined to drift away an hour or two, until the river bats awoke to flicker ghostlike over the song-dream waters of an early summer night. Here I would watch the early night flight of the Daubenton's Bat, feeding low over the river. This is my favourite bat, on account of its gracefulness over water. The gnats obliged by leaving me well alone, knowing perhaps that I was part of an idyll that could not be disturbed. Distantly, through eyes that were half-closed, I glimpsed a Stoat over the grassy slope, black-tipped tail, bounding into night-fall.

Part of me drifted downstream, a stranded Mayfly on early summer waters, as the resident cock Blackbird ascended to his chosen perch on a topmost Hawthorn spray. Choral evensong had begun, and promised to be as eternal as the singing river. He sang of Love, Joy, Beauty, Wonder and Awe, or so it seemed. Keats would have understood it. My role was to offer some form of blessing.

Dovedale, Derbyshire, 11ᵗʰ June.

> where the river's cleared its throat of storm
> and lies low, sucking stones, licking sky blue.

Chris Kinsey - *Hay*

Dog Days at Catfield Fen

Somewhere, away over the marshes, a belated curlew gave his long
whistling cry.

Arthur Ransome - *Coot Club*

Flat arable land fringed by dark legions of Alder, Birches, and
Oak; roofed by a sky so vast that heaven seems to be descending to earth, with towering cumulus clouds more actual than
land; secreted throughout by sinuous river valleys, broad and
shallow; masked and jungled by trees and vast beds of tall
Reed. This is how the Norfolk Broads seems to the newcomer.

Most visitors arrive exhausted, because it is a very long way
from anywhere else, probably because it wants to be. Distance
is everything here, in and around the meandering shallow
valleys of the Ant, Bure, and Thurne; and yet there is a near-
ness, a proximity to nature. Within this landscape our dislo-
cation from nature has been stalled, by nature itself.
Consequently, the Broads remain one of nature's last outposts
in England.

It is not perfect, of course. Vast swathes of land are dominated
by intensive agriculture, much of it arable; public access on
foot is grossly inadequate for a national park, especially along
the water's edge – so that, except by hired boat, one sees surprisingly little of the famous water. And the fen winds wreak
havoc when they blow, and blow they often do. Also, Broad-

land is one of those places where one feels the sun should always be shining, but it regularly doesn't; in fact, the whole of East Anglia is as much cloudscape as landscape.

The leafy lane westwards from Catfield village twists and turns and eventually bumps, in first gear, before petering out altogether close to a grumpy bungalow owned, it seems, by dangerous dogs (which have perhaps devoured their human carers). Imagine driving for several hours, only to be greeted by the Hound of the Baskervilles and its friends… But this is one of the few places in the Broads where those in the know can wander for a few hours in peace, at least during the week. One can penetrate far into the heartland of the Broads here.

I was fortunate: the trackway that runs southwards along the edge of the dyke (called a lode in Norfolk, a rhyne in Somerset and a sewer in East Sussex) had recently been topped, so that one didn't have to push continually through thigh-scouring Creeping Thistle stands and the ever-resurging *Phragmites* Reed. Better still, a friendly black Labrador, the antithesis of the hateable hounds up the lane, adopted me, and guided me all the way to the banks of the River Ant; a walk of a couple of hours at my pace. With glee, I hopped over a five-bar gate I could more easily have opened, or even circumnavigated – for no fence adjoined it, instead there was a welcoming gap. Ahead, wavering high in the sky, was the day's first Marsh Harrier, hanging, wings spread and angled upwards, flight feathers splayed, before veering off into a sky dotted with benign cirrocumulus clouds. Marsh Harriers scarcely seem of this world, let alone English.

To the right, westwards, was a large area of boggy Fen Sedge, shimmering with surface water and backed by dense, dark Alders; to the left, across the straight and narrow lode, a vast area of Reed bed, patches of which are periodically cut for thatching Reed. The lode itself, sunk below high banks of collapsing vegetation, was fringed with Yellow Flag, Great Water Dock, Hemlock Water Dropwort and populated with patches of yellow-flowering Common Bladderwort, white Frog-bit and assorted floating Pondweeds. Below the surface a whole host of Milfoils and Hornworts swirled and seethed. Best of all, the near end, where the water is stagnant, was choked with Water Soldier for a hundred metres or more, their spiked leaves reaching out of the water and finger-like towards the sky. This is water beetle heaven, and a place of pilgrimage for members of the gloriously eccentric Balfour-Browne Club who pursue and study Britain's 269 (or so) species of water beetle. I stared into the water's darkness awhile, contemplating rare water beetles such as the giant (4cm long) King Beetle *Dytiscus dimidiatus* and another rarity with the appealing name of *Hydrochus megaphallus*. For a while I gazed into the depths, pondering a mega-sized phallus.

Here and there, along the bank, patches of Bog Myrtle told me I was on acidic peat. A few sprays had been eaten back by Emperor moth larvae: those giant but cryptic plump green caterpillars you can fail to spot beneath your nose – unless you're colour blind, in which case they stand out like sore thumbs and can be spotted from many metres distance. There are, it seems, hidden advantages in some apparent disadvantages.

This is serious dragonfly country. Norfolk Hawkers, bronze-bodied and with bulging luminous-green eyes, patrolled low above the lode, occasionally intermingling and battling with the last of spring's blueish Hairy Dragonflies and that thug of a beast, the vibrant blue Emperor Dragonfly. All these mingled, in places, with pockets of male chaser dragonflies of several species. Chaos ensued, with much irate clicking of wings, testosterone, and rampant male egos. The females kept out of all this: a wary female Norfolk Hawker was curling her abdomen into the water to lay eggs on submerged Water Soldier leaves, a plant with which this Broadland dragonfly is strongly associated. Out of the maelstrom, also, small colonies of well-behaved damselflies were lying low: common Red-eyed Damselflies basked on floating Pondweed leaves, and at least three species of blue-bodied damselflies.

This is also grade-one hoverfly country, supporting many of the numerous species whose larvae are aquatic or semi-aquatic. Our scarcer insects are often abundant, or even super-abundant, where they do occur – none more so than the little black-and-orange wetland hoverfly *Tropidia scita*, which abounds here in such numbers that you can even hear its tiny wings whining. Golden, squat-shaped hoverflies, of the genus *Parhelophilus*, basked on leaves protruding over the water's edge, out of reach and consequently out of identification range: there are three species of them, all of which occur in the Broads. I obtained some excellent photographs of one male, but still couldn't positively identify it as you need to be able to see the inside of the hind leg, a part of the anatomy difficult to photograph.

And with the hoverflies come the soldier flies, those highly-coloured denizens of fens, swamps, and springheads. Most of them are so small that you only discover them when brushing waterside vegetation with a sweep net, but here at Catfield some of the commoner species are so profuse they are hard to ignore, notably the ubiquitous iridescent-green *Chloromyia formosa*, which has recently been given the vernacular name of the Broad Centurion. And then there are the biting flies, most of which hang out in the carr woodland or along the wood edge; out here, in the open you're relatively safe – failing that, rub Bog Myrtle leaves into exposed flesh: it's a natural insect-repellent and a useful get-you-by if in your hurry to get out on site you'd forgotten to apply the proper stuff – jungle juice – and left it in the car boot (which, of course, I had).

All this hits you within a hundred metres or so of the un-circumnavigated gate. Had I been diligently recording flora and fauna I would have been well into three figures by now. The dog ignored it all – panting at me to hurry up. "It gets better further on", he seemed to be saying.

It did. The day's first Swallowtail butterfly crossed the lode, a couple of hundred metres or so downstream, inspected a stand of Marsh Thistles, causing me to hurry towards it – only to cruise off and away over the Reeds. That's Swallowtails for you: unpredictable and aloof in the extreme. But no other British butterfly gets one's blood up more than the Swallowtail, except the mighty Purple Emperor. A scatter of Marsh Hog's Fennel plants along the Reed edge revealed a few Swallowtail eggs, dull orange, almost the size of a glass pinhead, on the under-sides of fronds standing proud of the surrounding vegetation. The British Swallowtail, long confined to the Broads, is wholly dependent on this rare wetland plant. The continental sub-species, on the other hand, is far more catholic in its foodplant choice and may have a more viable future in the long term here. It may well colonise England, and hybridise with our native subspecies – and in so doing engender an excellent conservation dilemma.

At the sharp corner, where the lode turns abruptly eastwards, lies a sunken pool surrounded by bushes. Damselflies danced, shimmied, and blushed; then from out of the bushes erupted a loud stream of coarse vindictive, ushered by a Cetti's Warbler. The catalytic explosion made me jump. No, on no account must Cetti-speak be translated into English. But at least I can identify it, whereas I struggle to separate out the songs of the Reed and Sedge warblers: the situation is scarcely helped by the fact that the two seldom occur together, preventing comparison, and are encountered mostly as partly-obscured shapes amongst wavering Reeds.

The black dog led me along a straight lode-side track bordered by plain Reed and Stinging Nettles. At one point he crossed the lode, by means of a ligger – a narrow black plank. I tried it: it wobbled badly, but by then I was half-way across and it was too late to turn back. Once over, the wobbly ligger proved to be the only way back to the land-of-the-living. In childhood and youth such challenges are readily overcome, but to the over-50s they are downright petrifying. Never again.

Sapphire-edged Small Tortoiseshell butterflies were emerging. You could see where their gregarious larvae had been feeding on sunlit Nettle patches – the distinctive protective webbing spun by the young larvae and leaves half-eaten by larger larvae were salient. Some Nettle patches still bore larvae, but these were the shiny and spiny black larvae of the Peacock, half-grown: they would emerge as handsome butterflies in mid-to-late July.   Another Swallowtail drifted briefly over the Reeds, again far away.

The track twists a little. On the left, now to the north, stands Middle Marsh Drainage Mill, a derelict brick-built windmill without sails – effectively a rural pepperpot. A Barn Owl was hunting over the Reeds close by. For some inexplicable reason, Barn Owls are far from nocturnal in Norfolk and are quite often seen out and about in broad daylight, even in the heat of the midsummer sun. This and other local idiosyncrasies are summarised by the expression 'Normal For Norfolk', which is abbreviated to N4N. A giant horsefly droned by, seeking man-flesh most probably.

Close by I came upon a loose posse of Swallowtails. They were males, aggregating around a flowering Bramble bush. They were seeking nectar but elected to squabble instead. The dog lay down in the shade. Eventually all bar one of them flew off; enabling me to admire the largest and brightest of our resident butterflies – one which makes you believe that the Mediterranean and the tropics exist within the blessed realm of Albion, at least in terms of wildlife. An old boy, he had lost his Swallow-like wing tails.

A little further on, a male and a female met up over a Thistle stand. She ascended, in an effort to reject his advances, then plummeted, almost to the ground, successfully shaking off her ardent suitor. She was already mated. Female Purple Emperors do something similar – dropping to the ground, bottoming out at the last second and sometimes causing the unwanted male to crash-land. Maybe those two gigantean butterflies have more in common than size?

The path, if such it was, narrowed, and disappeared into Alder carr, arriving on the banks of the River Ant at a place called, somewhat ingloriously, Mud Point. Here my over-heated companion waded into a shady pool and immersed himself in Duckweed, whilst I sat down to watch the slow-moving river and attract errant mosquitoes and the Common Deer Fly *Chrysops caecutiens*. The latter is particularly stupid: a large and obvious biting fly, with psychedelic eyes. The females fly around your head for several minutes before seeking to settle, and bite. There's no excuse for getting bitten by a Deer Fly, other than outdoor fornication, or ensuing slumber. Nothing else was happening, bar some quacking ducks.

A motor cruiser throbbed by, muddying the waters further and blasting out jazz. It reminded me of the Margoletta, crewed by the evil but ultimately hapless Hullabaloos in Arthur Ransome's *Coot Club* – a tale of daring teenagers and dastard grown-ups, further down Broadland's river system. Then a pair of twelve year olds sculled past me in a green rowing boat. Why they weren't suffering double French, double maths and double physics on such a splendid summer day I know not. Free as the river itself, they represented all that Ransome believed about childhood. He was convinced that freedom – freedom in nature, freedom to take risks, and freedom to learn through adventure – was essential to growing up. Furthermore he believed in the importance of childhood fantasy worlds, which parents should understand and nurture. What he would think of today's control-freak, risk-averse parenting beggars belief. He was not of this age. Who is?

I woke from such musings to see the diminutive figure of my friend, the black Labrador, sauntering off in the distance, back down the track. He was taking himself home. An hour or so later he greeted me, still somewhat covered in Duckweed, by the un-circumnavigated gate. I gave him my leftover sandwich. I had been delayed by Swallowtails basking in the evening sunshine, on the bare parts of the track where dredgings had been dumped sometime before, and by another basking, flat out, on Yellow Flag.

Such days do not merely linger on in our minds: they live on and actually develop within our souls, reaching a depth of meaning almost beyond our comprehension, for such emotions are rather beyond the medium of words – except for poets, which is why the poetic approach to nature is so relevant. I live for such days, and am determined to hunt down the words that describe them, so that they may be shared – and understood. But what matters more is that the love they generate becomes part of the spirit of the place, and dwells and grows there, somehow – and inveigles its way into other beings' lives.

Horsey, Norfolk, 25th June.

How beautiful the place is. Watch it hold
time still. I want you to tell me what this is,
this place at the back of beyond, in the sun
that retains its distance in a pale gold
mirror, minding its own brilliant business,
not in the habit of speaking to anyone.

George Szirtes - *Backwaters: Norfolk Fields*

Grey Knotts

Light & darkness coexisted in contiguous masses, & the earth & the
sky were but One!
Nature lived for us!

ST Coleridge - *Notebook, Keswick, circa 1800*

Mountains offer a serenity that no other landscape can offer, but only when they are in the right mood. Often their peaks are shrouded by mists and vapours, or worse – invisible. At times they sing with running waters, including from underground seepages and rills. Much of the time, though, it is just stinking wet. Our experience of mountains is ruled by weather, which the mountains themselves generate, like the banner clouds they sometimes unfurl from their peaks.

Crucially, up here, spring and summer are concertinaed together, within a magical six-week season, before autumn slowly descends. The best time to visit the Lakeland high fells is around or shortly after Midsummer Day, when they are at the height of their annual cycle of natural beauty. The litmus test is whether the valley farmers are making hay or not: if they haven't started, you are too early and will find the lingering vestiges of spring, and summer in its incipiency; if it's already been brought in you're too late, and the mountain grasses will have browned off, and autumn will be calling.

This year, I had first ascended Grey Knotts from Honister Pass in late May. It was Ascension Day, which perhaps explained why the weather was indifferent. The fell tops were clad still in the drab colours of winter, whereas the Oaks in the valleys below were in full leaf, and the Bluebells headily at peak. Up here, it felt like early March: the odd Violet was showing through, wan and pale, but even the Tormentil which abounds amongst the sheep-grazed Mat-grass turf was scarcely showing leaf, and had yet to form buds. Here and there, the Heath Rush was starting to produce black-and-yellow flowers, and some sheep-browsed Bilberry shoots were flushing amber-green; only the abundant Club Mosses, which make up much of the mountain vegetation, were making any effort to grow. The Sky-larks and Meadow Pipits, though, had better ideas, for as the sky brightened vaguely, and ceased to threaten, they took to the air, and to song. But then, as the weather changed its mind, and glowered, they fell silent again.

Then somehow the lightest of breezes started up, as if by grace of the mountains themselves, and the withered-and-weathered heads of last year's grasses began to waver. A ragged piece of blue sky appeared leeward of High Still; and from somewhere across Honister Pass, on the slopes of Dale Head, a Ring Ousel called, distantly. Closer by, more Larks and Pipits sang, and a Mountain Bumblebee droned lazily by. Then, sudden and all too briefly, a Wheatear broke into song, scratchily, like a Common Whitethroat, but singing in mountain-time.

Yes, I had been there the moment spring arrived. Now return-ing, as a high-summer anticyclone was building over Lakeland, I knew I was acting as a standard bearer for summer, even though it would prove all too ephemeral. You have to pace yourself carefully though, in such conditions, for it is easy to get over-excited and burn out on the fell tops in hot weather. One can probably die of ecstasy up here.

Ascending the old straight mining track, from Honister Slate Mine up to the heights of Fleetwith, joy filled every step. The secret, though, is to avoid breaking sweat – once sweat beads start to form, they continue for the day; and one is always carrying too much up a mountain, especially since fall-out from the Chernobyl nuclear disaster in 1986 rendered the fell top water undrinkable. Also, ascending a popular route makes one vulnerable to joining the race to the top which younger, less-experienced, walkers and the hyper-fit fell runners try to inflict on everyone else. Resist, resist: choose your own pace, and pause, and look around.

Today, the boulders were well-coloured, with the greys, whites, yellows, and luminous-greens of lichens. Soon the day's sun-shine and heat would reflect off them. In the boggy ground the Stag's Horn Club Moss had changed into diverse colours – greens, vibrant and otherwise, oranges, reds, and browns. Strands had grown into contorted shapes, often a metre long. A Herdwick lamb, coal black, bleated after its grey mother. Soon it too would become grey, with a buff-coloured muzzle.

Wheatears were nesting in a hole in the neatly-piled slate and rock spoil that makes up the old mining tram causeway. Wheatear nests are not easily found, except when the parent birds are ferrying moorland insects to the nest cavity – and even then it is virtually impossible to see into the nest. A torch is necessary, and a small mirror angled on the end of a short rod: more equipment.

Reaching the lower slopes of Grey Knotts, one of the main peaks on the south side of Honister, the sun became warm. Distant mountains were already beginning to haze – Skiddaw and Blencathra to the east, Scafell Pike to the south. But here the ground was embroidered with yellow Tormentil flowers and, in places, silvered with Lady's-mantle foliage. Some base-rich flushes, heavily favoured and grazed down by sheep, were pink with drifts of Wild Thyme flowers. The flowers of spring and high summer were blooming together here.

And everywhere, below the cry of the Meadow Pipits, the montane insects. The moorland cranefly *Tipula subnodicornis* was hatching in numbers. This blue-grey daddy-long-legs is a vital part of the food chain up here, in both its larval and adult stages: ask the Meadow Pipits and the Wheatears, for which the fly is a staple diet. Mountain bumblebees were out and about: notably a black-yellow-and-white one that dashed about frenetically, visiting Tormentil flowers and proving impossible to photograph – it might have been the rare north-ern mountain species *Bombus cryptarum*, the Cryptic Bumble-bee, which can only be determined by DNA analysis. Someone else can do that. More reasonably, the handsome yellow-black-and-flame-red Bilberry Bumblebee *B. monticola* was seeking out its favoured Bilberry flowers. A Wood Tiger moth darted out of a tussock in front of me and zig-zagged away at speed, low, before disappearing into the grasses. This is a scarce day-flying moth, bright in the colours of a butterfly, which most naturalists know from the southern chalk downs, yet it also occurs quite freely on the Lakeland high fells, even 600 metres up. Down-looker Flies, the snipe fly *Rhagio scol-opaceus*, settled on every boulder, and on every pausing fell walker – most of whom reacted adversely, expecting this harmless percher to bite. Everywhere, around the moorland pools, the tiny iridescent-bodied Dolichopodid flies – known, somewhat belittlingly, as long-legged flies: for it is their metallic green and blue body hues that make them memorable.

And here and there, cruising low over the ground before suddenly crashing, moth-like, into a tussock, the sooty wings of the Mountain Ringlet butterfly. These were males, searching ardently for females. They were attracted to anything brown, even cast-off pieces of Herdwick wool and sheep dung, in the hope that it might be a virgin female. Insulted perhaps, the females were lying low, gestating their eggs in warm pockets between Mat-grass tussocks, and avoiding the males: they were already mated, and not in need of male services. In their own time they would fly short distances, crash and then crawl, to lay their pale yellow eggs on the blades of Mat-grass or Sheep's-fescue grass, then skulk once more, wings closed, dense black hairs on their northern faces, in the tussocks. All too suddenly the colony ended, though the grassland habitat seemed unchanged.

It was time to wander on, further up and further in, southwards and sunwards; up towards Brandreth and Green Gable, pausing to gaze down into Buttermere, way down to the right, which was starting to shimmer, and then into the lonesome wildlands of Ennerdale, with its ever-changing, braided, pebble-strewn river. The mountains were calling, and would not be denied. Some journeys need no end.

Coleridge, that great metaphysician of the Lakeland fells, argues that the poet is a metaphysician who actively engages with nature, who ventures outside of himself to hunt down 'the otherness of being' (originally paraphrased by Richard Holmes in *Coleridge: Early Visions*, 1989.) What, then, is the naturalist, the lover of wild nature? Few have even posed that question. Perhaps naturalists have little notion of their true identity, for their passion has been overtaken by, and wholly subsumed into, science – and perhaps rather narrowed down. The Lake District is now in the process of being designated a World Heritage Site, partly (and rightly) for its links with the Picturesque and Romantic movements, and partly (and wrongly) for its significance as a cultural, sheep-farmed land-scape. That combination is utterly untenable – ask any poet, or ecologist – and will result in bureaucratic chaos through which the Lakeland farmers will (perhaps rightly) drive a coach and horses. Above all, Lakeland is a spiritual landscape; and by that I mean a landscape which appeals so strongly to the human soul, spirit, psyche (call it what you will) that it forces us to realise and accept the vast metaphysical aspect of our existence. Few have understood this. Coleridge did.

Brandreth, Lake District, 24th June.

NO ONE CAN LEAP OVER HIS SHADOW;
POETS LEAP OVER DEATH

ST Coleridge - *Notebook, autumn 1801*

A272 Revisited

> White clouds ranged even and fair as new-mown hay;
> The heat, the stir, the sublime vacancy
> Of sky and meadow and forest and my own heart
>
> Edward Thomas - *The Glory*

June was coming home, albeit under a chaotic sky of varied altocumulus clouds – suggesting that weather was seeking to arrive from different directions all at once, to cause chaos. The mixture of low-, medium-, and high-level clouds of differing shape, pattern, intensity, and density suggested inclement weather to come. Amongst it all, though, there were patches of blue, offering hope. Such is June, so often.

Summer's Hop trails were starting to twist and turn their way up the windblown Scots Pines that lean drunkenly where the old A3 meets the backbone of the Purple Emperor's kingdom, the A272, near Petersfield. A pair of Spotted Flycatchers were mopping up hoverflies. They, and their ancestors, had bred there for many a long year. I know, for I used to hitchhike from the adjoining layby.

Further along, eastwards, Buttercup time was ending at Durleighmarsh Farm, as a carload of out-and-abouters turned in there for afternoon tea at Alexandra's Kitchen (a yurt in all probability). Placards erected randomly along the verges were advertising, and advocating, Rogate Village Fete and Milland Rural Fair, both to be staged on the same day. Rogate itself was placard free, offering instead rampant rambler roses – flame orange and dark red – and some up-eaving House Martins. Just past the village, at Terwick, lingering clumps of purple and pink Lupins gave up their final flower spikes of the year. There was a time, not long ago, when Lupins would be at peak in mid-June, but in the modern era of mild winters and early springs they are at their end by then.

All along, were lingering patches of Cow Parsley, clumps of Red Campion and the deep-red clover of June – at least where the flail cutter had not already ventured: to masticate spring into myriad decomposing fragments, each vehemently des-ecrated. (The people who operate those machines must seri-ously hate nature, as some people do.) The only spring colours remaining were the still-fresh green of Bracken fronds and some late Ash trees that were belatedly unfurling their final leaves.

At Cumbers Hill, a Copper Beech was telling of colours to come in that unnameable season which arrives when summer's wonder ends; whilst in the ancient woodland copse on the hill's eastern side, the Bluebell leaves had yellowed and flat-tened – soon they would vanish altogether. Sallow seed was drifting across the road, here and further along, with the same intent and purpose of Dandelion seed, now gone by, or of thistle down, yet to come.

Midhurst had drifted off into a timeless slumber, its High Street clock stuck at 4 o'clock: teatime. Eastwards, the tumbling water in Easebourne weir pool was telling of the start of the coarse fishing season, shortly to come, on 16th June. Elsewhere in Easebourne, the Lilac blossom was ended, and had turned brown; a cock Blackbird preened in a Judas tree, bearing the last of its almost synthetic Baker-Miller-Pink blossom; petals lay on the pavement, unnoticed by people in passing vehicles, save one.

Beyond, on the Lower Greensand ridge, where the later Sweet Chestnuts still held light green leaves, the Hawthorns were dropping discoloured petals. Their flowers were being replaced by June's Elder blossom and, at Tillington, a patch of early Bramble blossom.

The 200-year-old wall around Petworth Park offered Mexican Daisies, trails of purple Ivy-leaved Toadflax and some yellow *Hieraciums*. The town itself was winging with Swifts and chundering with caravanettes and lorries, to which the Swifts were rightfully oblivious.

At Foxhill, the crest of the wooded ridge east of Petworth, the world suddenly changed: a yellow Laburnum – the last of spring's blossoming shrubs to flower – leant over the road, punctuating the start of a tunnel of Oaks already darkening towards July, and framing a profound vista into the wooded Weald ahead. Here the song-dream became loud and clear, and almost tangible. It ran one step ahead, hurtling towards the heart of summer, which we belittle somewhat by using the humble term 'midsummer' – 'peaksummer' would be more appropriate, 'zenithsummer' better still. I had to follow, even more intensely. It was leading me towards Knepp Wildland, deep in the very heartland of West Sussex.

The other side of Billingshurst, the horsebox which had been blocking the view since Hawkhurst Court – and chaining me to a reality of which I do not want to be part – turned off to Itchingfield (and damnation by eternal itching, one hoped); and the way ahead, deep into the heart of summer, was wondrously exposed.

Knepp Wildland is a new creation: nature's. Disillusioned with the difficulties of mainstream farming on low-grade agricultural land, on heavy clay, and inspired by the reasoning of the wise Dutch ecologist Frans Vera into the nature of the wildwood, Europe's original land cover, the landowner here took the immensely bold decision to take almost his entire 1,400 hectares (3,500 acres) estate out of modern arable and dairy farming, remove all internal fences, and allow free-ranging large herbivores to determine the land's future. The theory is simple: in the wild state, large herbivores are primary drivers of the ecology of landscapes; so, remove as much human interference as possible, erect a strong perimeter fence, intro-duce old-breed cattle (Longhorns, as analogues of the wild-wood's aurochs), deer (Red and Fallow), some Tamworth pigs (to replicate the ground disturbance produced by wild boar), and a few Exmoor ponies to buzz off surplus grass; then stand back and accept the beneficial changes that this naturalistic grazing system produces. But it is so difficult for us to stand back: we so earnestly desire to be in control, and determine what we term the 'balance of nature' – by being in control of it. Of course, there is no balance of nature, for nature is a dynamism, a vortex of change...

The Knepp land is still agriculturally productive, generating some £120,000 of meat sales a year, primarily in the form of top-grade organic beef – and needs to remain as 'agricultural land' for taxation and legal purposes. The estate's main income, subsidies apart, comes from farm buildings which have been converted into business lets, and commercial rents from cottages formerly occupied by agricultural workers; so, the money comes from buildings, not from exploiting the land. In effect, the main block of Knepp Wildland, some 470 hectares (1,128 acres), is developing into a modern version of the wood pasture system that dominated much of Britain (lowland and upland) for centuries, before the Agricultural and Industrial Revolutions.

Crucially, Knepp Wildland is about mending, the mending of the immense damage done to the land, nature, and our relationship with it, and to spirit of place, by the EU's Common Agricultural Policy and the global agri-chemical-mechanical industry. It is as dramatic, and wonderful, as summer's own mending – of time damaged by winter. I had to be there, where the Lesser Whitethroats had gathered in numbers along the outgrown Blackthorn hedges, and as the Cuckoos were ceasing to call. For Knepp itself is a calling.

Shepherd's hut at Knepp Wildland, 18th June.

Man comes and tills the field and lies beneath,
And after many a summer dies the swan.

Alfred, Lord Tennyson - *Tithonus*

Musings through Madgeland Wood

> glories infinite,
> Haunt us till they become a cheering light
> Unto our souls, and bound to us so fast,
> That, whether there be shine, or gloom o'ercast,
> They always must be with us, or we die.

John Keats - *A Thing of Beauty is a Joy Forever* in *Endymion*

Throughout the journey through spring and into summer there are features – of time, place and circumstance – which appear suddenly as if from nowhere, and surprise us. This is because there is so much happening within the hurried, almost frenetic, temporal sequence that it is hard to anticipate anything. Even the first Field Rose or Honeysuckle bloom can come as a surprise, despite them being major emblems of Summer. Perhaps we do not live close enough to nature to anticipate, or foresee?

I was walking a wood I've known for nigh on 50 years: looking for the Oaks' first Lammas shoots of reds, or orange, and for the blueing of the Oak foliage that heralds Purple Emperor time – when suddenly I was arrested by Aspen leaves dancing in air which otherwise appeared still. The late morning seemed breathless, yet the Aspens were trembling. Then, to more amazement, and again quite without warning, though none was needed, the summer's first flock of young Titmice came tumbling and calling through the tree canopy: a bevy of eight or nine Long-tailed Tits. It seemed all so unexpected, yet so utterly proper. Nature continually catches us unaware, and we continually take it for granted.

It is easy to fail to anticipate the appearance of the more humble beings – like the first of the see-through hoverflies that waver above you along woodland rides from mid-June to early August (and sometimes before, or beyond, for timing of appearance is not fixed, especially with insects). It has no vernacular name, but goes by the Latin name of *Volucella pellucens* – with the two names respectively referring to hovering and the pellucid state of transparency. It breeds as a scavenger in bumblebee nests, but is a rotten bumblebee mimic, though it does have an impressive see-through abdomen.

This is one of the thousand-and-one creatures we readily encounter, and ignore – because we don't know its name, or anything much about it. Our world is factual, too factual perhaps; while nature's world is concerned with interconnectivities, relationships, and associations. Our knowledge of these myriad interconnections is grossly inadequate. It's in the 'too complex / don't go there' box.

We are obsessed with names – but the naming of something is only a small part of the understanding of it, and not necessarily the beginning. Instead, seek the essence of the thing, the meaning beyond the words, the relationship with the plant or animal, and its myriad associates, the meaning beyond the meaning of words, the ineffable, the numinous. Look and thou shalt find; ask and it will be revealed unto you.

Connecting with nature is about falling back in love with the natural world, and at the same time letting it fall in love with you. It is a two-way process. Who knows, one day nature might take the lead and decide to reconnect with us...

Madgeland Wood, West Sussex, 19ᵗʰ June.

Woods fill you deep as your first breaths.
Before the great fritillaries
float flame across slow sunlit paths
you taste the trees,

their reek of root, their restless crests,
scarce water trickled through the lime,
high branch which broke from heatwaves' stress,
hard rings of time.

Alison Brackenbury - *Why*

Emperor Time

What I love shall come like visitant of air,
Safe in secret power…

Emily Brontë - *The Visionary*

For some days my watch battery had been faltering, but in late June one is too busy to visit a shop to change it. The watch stopped altogether on the day the year's first Purple Emperor took to the high summer skies, 27th June. I discarded the watch amongst the dust balls gyrating mindlessly around on the floor of the shepherd's hut at Knepp Wildland. It didn't matter anymore: I had entered Emperor Time – a truer reality than the one we exist, or perhaps discombobulate, within. Indeed, for the most part, our lives are acted out in mime, so as to meet expectations, and we become our true selves only on one-off occasions such as once-in-a-lifetime holidays, and honeymoons. We need more and truer honeymoons: natural honeymoons. The Purple Emperor season offers five or six weeks of rapturous honeymoon with nature.

Unprotected by the watchstrap, the band of pale skin quickly reddened in hot sunshine, as June flamed its way out. At any other time of year it would have chaffed, even caused sleepless nights. Now, it went unnoticed. Mind was acting over matter.

After a morning of evaporating cloud, the year's first Purple Emperor flickered momentarily, a dark spectre high up on the leeward side of a mighty Oak at the northern end of the ancient green lane that bisects Knepp Wildland. Only a flicker, pecking at some mystery object of interest, before drifting off over the Sallow thickets beyond, but enough: the ecstasy of the year had begun. Radiance filled the air, and me.

A short while later, further down the rambling Oak-lined lane, another Emperor appeared, again from out of nowhere – flying at speed straight at my head, aiming at the eyes. I froze, as one should when experiencing divine visitation. At the last millisecond this alarmingly brave – or berserk – insect veered away, up and over the bushes, having flashed and clicked his iridescent indigo and electric-blue wings in front of my eyes. Nothing else sounds like the clicking of the Emperor's wings – it is diagnostic – but oh, to understand the language! "I know you," I shouted after it, spontaneously using Edward Thomas's words, "You are light as dreams…"

That's early season Emperors for you, brazen and inquisitive in the extreme: pioneer males, lording it, capable of trying anything, grand masters of experiential trickery, and gloriously amoral. Not for nothing did the Victorian butterfly collectors call this regal insect 'The Emperor of the Woods', 'His Imperial Majesty', 'The Monarch of all the Butterflies' and, even, 'The Lord of the Forest'. But these powerful epithets sell HIM short: He is the High Spirit of the Midsummer Trees, the One of Whom the Nightingale Sings, the One for Whom Summer Comes. And he knows it.

To its devotees, the Purple Emperor season is utterly liberating. Not only does it blast normal human-time away but it also instils a greater reality than the one in which we lie, so harshly trapped within. And we are entrapped, within a material world of our own manufacture, dominated by money and increasingly convoluted systems, spiritually suppressed and driven by the power of assumption. We have enslaved ourselves in bureaucratic processes; and sold our souls to materialism, the thing the Old and New Testaments both warn us most against. The Emperor dismisses all that baggage with a single flick of his wings: only a flicker. The rapturous sense of freedom is overwhelming, and monumentally real.

Break free! Choose love, joy, beauty, wonder and awe (even though the latter is at times alarming). Cast your watch into the dust, and then your cash cards. Transcend the material world, and look on the Purple side of life. I'm not advocating some shallow momentary experience, soon to be forgotten, a two-dimensional television experience; but a total immersion into the real world – of nature: the world of wonderment, at least for a few heady weeks beneath the midsummer trees. Nature offers us life-changing experiences; but it is not a superficial tourist experience, it is our place of rightful belonging.

Knepp Wildland, West Sussex, 27th June.

Merrily, merrily, shall I live now…

William Shakespeare - *Ariel's Song* in *The Tempest*

Countisbury Cliffs

Like a Poet hidden
In the light of thought,
Singing hymns unbidden

Percy Bysshe Shelley - *To a Skylark*

The tiny hamlet of Countisbury nestles over the brow from the steep north-facing cliffs of the Exmoor coast. It consists of an old, thick-walled roadside inn dominated by dogs and chips-with-everything, a scatter of farmsteads – now mainly holiday homes or National Trust holiday cottages, and a church secreted away in a fold in time and place. Sheep run free in the road, oblivious of tourist traffic. Nothing much has happened here, at least not since an invading force of Danes was wiped out during a hillfort battle in 878 AD. Yet people, and time, have passed through Countisbury, for the A39 coast road plunges through it, beginning its lengthy descent into the mouth of the Lyn Valley. It is a place of Meadow Pipits, Stonechats, Gorse clumps, sheep-grazed Bent grasses, low cloud and sea fog. To the south, downslope, hidden away amongst steep-sided Sessile Oak woods, is the magical East Lyn River, a place of singing, laughing, and dancing waters which might make you believe in dryads and hamadryads if you don't already.

A mile or so downhill to the west, lies sleepy Lynmouth. Here the young poet Percy Bysshe Shelley honeymooned with his 17-year-old bride Harriet back in the summer of 1812. He worked on *Queen Mab*, a poetic rant celebrating free love, atheism, republicanism, vegetarianism, and several other -isms scarcely relevant to us today, and his infamous and gloriously naive *Declaration of Rights*. The latter was deemed seditious, and the young Shelleys were closely watched.

Harriet wrote of the place: 'It seems more like a fairy scene … as if nature had intended that this place should be so romantic.' Her idyll was short lived, however. The radical, intellectual, but deeply impoverished Godwin family visited, including a daughter, Mary: she would become Mary Shelley after Percy Bysshe abandoned Harriet for her, while Mary's stepsister Claire would become one of Lord Byron's ill-fated mistresses.

A few blissful, if bizarre, months ended farcically, when the Shelleys were forced to flee across the water to Wales, after Percy Bysshe (who Harriet called 'Bysshe') had launched copies of his *Declaration* into the Bristol Channel in corked bottles, and his servant, an illiterate Irishman, had been jailed after being found in possession of this rebellious document. Four years later, at the end of December, poor Harriet drowned herself in the Serpentine, in an advanced state of pregnancy – the paternity of which is still being debated.

Earlier, in November 1797, the Wordsworths, Coleridge, and William Hazlitt had visited Lynton and Lynmouth. Coleridge and the Wordsworths were so struck by the place that they considered moving there; Coleridge being impressed not least by the Valley of the Rocks and by the coastline's 'august cliffs', fluctuating weather, and spectacular sunsets.

They would have passed through Countisbury, where the

church of St John The Evangelist had recently been enlarged. The church was extended again in 1846, and would have been the dominant presence in the community during the Victorian era, outrivalling the pub and attracting worshippers to brave the long ascent up from Lynmouth. Its churchyard is full of deceased Victorians, though their headstones are now lopsided and encrusted with lichens, yellow and grey.

Today, on a high summer evening, this church offers something very different to the incense and conditioned responses of Victorian Eucharist. Yes, the Ten Commandments are still inscribed on either side of the altar, it seems in dried blood, but they hold no fears to us – we wouldn't dream of doing nearly all those errant things anyway. This building now offers something radically different.

The church is kept open, at least to those able and bold enough to put their shoulder to its heavy, jammed, iron-studded door and push hard – having first apologised to the Swallows nesting in the porch (five featherless young, gapes wide open, anxious parents twittering outside; second brood anticipated). The door sticks, wilfully. One enters in hushed silence, as silence once more reasserts itself. Those of us who have experienced the contemplative worship of The Religious Society of Friends (Quakers) will not be put off by this omnipotent silence; but to people not used to such awe-inspiring silence this church may at first seem frightening. They expect noise, bustle, and church servicing, or even a meet-and-greet volunteer steward eager to explain the stained-glass windows or reredos. But when we move out, nature moves in.

My heart leapt for joy: everywhere there was dust, spiders'

webs, moulds and slime moulds; dead Cluster Flies, craneflies (three species), queen wasps (two species), a Map-winged Swift moth (larvae feed on Bracken roots) and ladybirds (mainly Harlequin) in the windows; and droppings that proved to have been deposited by Lesser Horseshoe Bats. Church mice had been hard at work here. A Jenny Wren had inveigled its way into the chancel, and told me off in no uncertain terms: these were its spiders, not mine, and I was not to touch them; and I was to leave now and never come back. I chose to stay. The Wren left discretely.

The place was clearly loved, deeply, and cared for, subtly: window recesses had been decorated with wreaths of Laurel, Yew and Ivy; there were everlasting flowers in everlasting vases, and on the simple wooden altar, a plain wooden cross and wooden candlesticks. The inner walls themselves feel prayer-soaked. Many of those prayers would have been for fine harvest weather or an end to the prevailing rains, for it is only recently that our lives have ceased to be dominated by weather.

Outside there were Jackdaws, who had long owned the church tower, a late warble from a cock Blackbird, flitting Swallows, spiralling seagulls and Swifts, and the cries of lost-and-found lambs. The churchyard would appear unkempt to some: consisting in this summer of rampant vegetation growth of Yorkshire Fog Grass, which had grown so tall in the rains that it had started to flop before it had dropped its pollen, equally tall Common Sorrel – and little else. The vestiges of spring flowers were visible underneath this mattress of dense out-grown vegetation.

Past a lone Scots Pine, through which soughs Shelley's wild west wind, the breath of autumn's being, and out through the back gate, the path leads quickly on to the open cliff top. There, twisted stems of wind-blown Foxgloves grow on the lee-ward edges of clumps of European Gorse, and Stonechats chide us – probably using foul and abusive language. There, one can watch sunsets a-plenty, each one different; or thunderstorms rampaging up the Bristol Channel – as Coleridge once did (he ran amok in it); or simply be enwreathed in low cloud, and listen out for the Raven's call.

Countisbury Church seems to have become a church of the outdoors – a Church of the Risen Christ in Nature, a Church of Christ in Creation, a Church of the Garden of Gethsemane. Some of us may need such a church, and may even recognise that we do. So many of us are seekers, in need of an outer church which can attract us in from off the wind-blown, storm-tossed hillside – without forcing us to believe in religious dogma or religiosity, let alone being forced to sing *New every morning is the love* on bitter February mornings, pray earnestly for the Royal Family, or confess to having sinned and been an abomination in the face of the Lord. An outer, outdoor-orientated church might help those of us who, like Thomas Paine (1737-1809) – a founding father of the United States and author of the *Rights of Man* – have made their own minds their inner church. Above all, a church is needed where people can better understand our relationship with nature, not least because nature was formerly called Creation; nowadays it's called biodiversity and the environment, which probably indicates how far our relationship with nature has deteriorated. An entry point for considerations here could be the Franciscan mantra: 'through Christ all things become our brother and our sister' (if you find that difficult, then start with the 'all things become our brother and our sister' bit; or substitute Love for Christ, they're arguably synonyms anyway). Shelley would hate this, and would launch an armada of bottled vitriol into the Bristol Channel.

The main use of old churches in this day and age is probably not for services (except on Christmas Eve when the C of E comes wondrously to life), or even to bring believers together, but to provide points of hope in the landscape, urban and rural. The buildings themselves do that, assisted by bats, Jackdaws, Swifts, etc. The spires and towers dotted about in town and country are beacons of hope, still points in our fast-turning world, touchstones with our spiritual life. And we are spiritual beings, existing within a spiritual landscape. CS Lewis is, wrongly, attributed with the saying: 'You do not have a soul. You *are* a soul. You *have* a body.' (The statement seems to originate from the Christian mystic George MacDonald, 1824-1905, who pioneered adult fantasy-writing with *Phantastes*, 1858, and *Lilith*, 1895). Whatever, our love for both nature and places is spiritual, deeply so; it's just that many of us struggle to recognise or accept this – perhaps because the concept is difficult to articulate and, perhaps, because our understanding of nature has become heavily-dominated by scientific thinking.

All this, and much more, started to become evident during an hour of peaceful contemplation in Countisbury Church one high-summer eventide, before a lone holidaymaker entered, noisily, and mutual embarrassment descended.

Countisbury, Devon, 30th June.

He prayeth best who loveth best
All things both great and small;
For the dear God who loveth us
He made and loveth all.

ST Coleridge - *The Rime of the Ancient Mariner*,
near Porlock, Exmoor coast, 1797

The People of Purple Persuasion

The people who walked in darkness have seen a great light…

Isaiah 9:2

The day promised glory, from the moment the sun began to burn off wisps of mist that had wavered above a short-lasting dew, outside the shepherd's hut. Although it was the start of July the Blackbirds were still singing, the knock-on from a late spring. They were on their second or third broods. But they weren't singing of spring anymore, but of the pinnacle of summer's glory – the Purple Emperor season – and it seemed that this was what they'd been ideating all spring and summer, such was the depth of their vocal passion. Soon they would fall silent, and go into tumultuous moult. Henry (or was it Henrietta), the tame Robin in the Knepp Wildlands's glampsite kitchen was already moulting but was still coming in to the open-air kitchen in search of his (or her, the sexes are inseparable) favourite repast of butter and oat flakes. It was time to get out on site.

Sedge Warblers were scratching and chattering amongst the Reed beds along the fringes of the old hammer pond, the surface of which was carmine-pink with the plumes of Amphibious Bistort. The birds would become quiescent as the day's heat developed. In early July, birdsong gives way to a different music in the air – the subtle but vibrant hum of myriad insects. By eight o'clock the insects were fully active, all save one who roosts high in the Oak tops and begins his day after a lengthy morning bask. His name is *Apatura iris*, named after the rainbow-winged messenger of the Gods, and he is Emperor of the Woods, and more, and he knows it.

The land to the north of the hammer pond is a former field called 27 Acres. This and its neighbouring fields were under intensive cereal production into the early 2000s, and were all but devoid of earthworms. After the 2004 cereal harvest 27 Acres was left fallow. The following May, seed drifted down from the scatter of Sallow trees along the edge of the lagg (a Sussex term for damp shallow valley), upstream of the hammer pond, to colonise the abandoned field. Other tree and shrub species moved in too, notably Hawthorn and Pedunculate Oak, though the Sallows, being fast-growing pioneer colonis-ers of bare clay, made all the early headway. Cattle and deer browse heavily on them in early spring, before the grass starts to grow. A large number of insects depend on them, plus their predators and parasites.

No one expected the Purple Emperor butterfly to move in quickly here, let alone establish one of the strongest popula-tions recorded in British entomological history. But it did, rapidly, colonising from a modest population in Southwater Forest a mile or so to the north, across the A272, at Dragon's Green. By 2013, 27 Acres and the surrounding ex-arable fields of Rainbow, Oak Field, Honeypools Barn and Woggs Bottom had developed vast jungles of dense, drawn-up Sallows – hybrids between the broad-leaved Goat Willow *Salix caprea* and the narrow-leaved Rusty Sallow *S. cinerea oleifolia*. These hybrids are known as *Salix x reichardtii* or, more simply, as Reichardtii hybrids. They are immensely variable, and surpris-ingly inconsistent from year to year. The Emperor breeds on them, favouring the larger, broader-leaved varieties.

Having spent ten months as creepy-crawly caterpillars, five of those in hibernation (or diapause), on Sallow stems where they are mercilessly predated by marauding Titmice; and then having suffered the traumas of pupation, followed by hanging upside-down from a leaf for three weeks, as vulnerable pupae: these guys want some action – and they have the wing-power to indulge themselves, ruthlessly. The males are obsessed with sex and a desire for world domination. Theirs, it seems, are the July skies.

The day's first Emperor appeared just before 9am: the size of a small bat – slowly and gracefully searching the crown of an Oak tree, before flushing out a small Purple Hairstreak which he duly sent packing. Ten minutes later they were all up and about – males ceaselessly patrolling the canopy of the Oaks that lined the old field hedges, meeting and greeting each other, in not the politest of ways, and bickering. Perhaps they had been roused, or wound up, by the hammer pond's Marsh Frogs, which had broken into full eructation, punctuating the calm of a warming July morning with croaks and clammerings, which sounded distinctly un-British. This is a non-native species which has spread far and wide in South-East England, and reached Knepp about a decade ago. But, after ten minutes, during which it sounded as if the lake itself was in the process of lift-off, the frogs fell silent – as suddenly as they had started up.

By now the Emperors were dropping down from the Oaks, to search the Sallow thicket tops for emerging females. This they do during the first half of the insect's flight season, primarily during the mid-to-late morning period, until the females are all out and mated. Then they suddenly stop this behaviour, and take the mornings off. Today, they were anticipating a size-able emergence of virgin females, and were frenetically search-ing the Sallow tops and surrounding wood edges. Inquisitive and belligerent, these guys will try anything. One chased off a hapless Chiffchaff at speed, another pitched in to a flock of young Titmice, and eviscerated them from the Sallows (exorcising some serious stored resentment), whilst a third took on and pursued a male Emperor dragonfly, out hunting over the thickets, in a spectacular demonstration of thug attacking thug.

Moving out of 27 Acres into Rainbow, you knew it was early July for the first Gatekeeper butterflies were emerging there. These are climate change deniers, like Swifts and Ash trees: appearing almost nationally on July 3rd, 4th, or 5th each year. Fortunately for them, they fly well below the avenging Em-peror's radar, amongst twisted grasses and budding Fleabane, along the scrub edge.

At the west end of Rainbow is a long stand of dense Sallow saplings, maybe 350 metres long by 20 metres wide, and backed to the west by tall Oaks. Here, half a dozen Emperor males were systematically searching the Sallow jungle, flying in and out of the tops, looking not just for females in need of male attention but for pupae containing females: males were repeatedly seen pecking at the same spots in the Sallow canopy. Suddenly, a male flushed out a receptive female, and pursued her graciously up to the Oak tops. Their follow-my-leader courtship flight is short and sweet. They normally mate, comatose and wings folded, end to end, for four solid hours –

during which time they are pestered by passing males, seeking to muscle in. Later, a mated female was intercepted by an amorous male. She immediately faltered her wings and dropped, literally, to the ground, in an effort to shake him off; at the last millisecond she bottomed out, leaving her suitor floundering in the Brambles, before rising up and taking it out on the first flying object he encountered – an innocent and naïve Meadow Brown.

From noon onwards, Emperor males, who have been unsuccessful in the bushes, rise up to the Oak canopy where they establish territories in prominent but sheltered canopy gaps, out of the wind's turbulence. There they wait in hope, for the arrival of any late-emerging females, or virgin females who had somehow remained undetected in the Sallows. These guys are full of testosterone and seriously worked up. They become highly belligerent towards anything that enters their airspace, other than an Empress – butterflies, other winged insects, and birds, small, medium, or large (the Knepp hit list here is impressive: 'small' includes Lesser Spotted, or more aptly, splatted Woodpecker, 'medium' includes Turtle Dove, and 'large' is spearheaded by Red Kite). But most of all they dislike each other, acutely.

Two males met over a gap between two Oaks, circled each other three times, spitting fire and brimstone, before one chased the other off and away, upwards and out of sight, at speeds only falcons, Swifts and Swallows are supposed to match. One returned, then the other; and it all began again. Then a Collared Dove passed through: which they both pursued, spouting vitriol.

Further down the Oak-lined green lane, a posse of three males ganged up on a not-so-innocent Spotted Flycatcher, and traumatised it, forcing it to change territory. Five minutes further along, a minor sap bleed, high on the underside of an Oak bough, had attracted two Emperors, a Comma, a Red Admiral, and a couple of Hornets to feed. At first the Hornets were mildly tolerated, merely warned off by powerful wing flicking. But soon, enough was deemed more-than-enough, and the Hornets were duly seen off. If Hornets are orcs, Purple Emperors are uruk-hai. This predilection for Oak sap may explain why the Emperors of Knepp, and Sussex generally, are more violent and more prone to bizarre behaviour than their cousins elsewhere: Sussex is a land of veteran Oaks, and veteran Oaks produce copious sap bleeds – the Sussex Emperors feed primarily on fermenting sap: and are perhaps seriously plastered.

It was mid-afternoon now, and the Emperors were quietening down – taking a siesta before becoming active again in the early evening. Though then their activity would become increasingly localised – confined to foliage bowls in the Oak canopy where the sun's rays linger longest. More Emperors, and a giant Empress herself, came in to feed on the sap run. Some lingered there till well past seven o'clock, with one lasting till after 8pm. Eventually, they departed to roost high in the Oak sprays, tangling with Purple Hairstreaks indulging in their evening flight.

At Knepp Wildland the favoured Emperor territories have been mapped and named. Some modestly so, after the more despicable Roman emperors – Caligula, Commodus and Nero; many after the more aggressive side of modern Britain – Skinhead Alley, Mindless Violence and The Serial Offenders Institute; a few after lascivious intent – Lord Byron, The Benny Hill Show and Lady Chatterley's Lover. Around this glorious and inglorious show, which lasts the better part of a month, has grown perhaps the most zany but riveting wildlife tourism enterprise in the country, run by the estate during the key weekends of the Emperor season. Sparring Emperors are saluted, and often cheered on. Drinks are raised to them. Purple Emperor devotees have been known to roam around with their trousers on their heads – occasionally starting the day in such mode, and deteriorating from there. Never mind rutting stags (or rutting Feral Goats, which are more badly behaved than stags), this is the Purple Emperor rut, and it deserves its own hour-long edition of *Planet Earth*, if not an entire series: indeed, viewers, you are being sold woefully short by the BBC Natural History Unit, for the best action's right on your doorstep. Without doubt, this butterfly boasts the biggest ego in the British Isles.

But ascend further, ascend deeper, into the glorious yet amoral world of nature. Here at Knepp Wildland, as the year's acorn crop swells, caterpillar frass drops softly to the ground, Syrphid flies hover and hum in dappled shade, Swifts and Martins fly high in a cirrostratus sky, and the song-dream is almost tangible: we can leave all our baggage behind – it doesn't matter anymore; it scales right back, to insignificance. I mean, the whole material, fiscal, and peer-pressure world that stifles us; and most of all, the mental and spiritual suppression generated by the all-powerful, soul-destroying industries of politics, business systems, managerialism and human resources. Beneath the Emperor's mighty trees, beneath his universalist skies, all of that dross pales into irrelevance, as if it had never been, losing its grip on reality; and rightly so, because it is not properly real, but an imposition. Emperoring – as its devotees call it, does not merely set us free, it restores us to our natural state – places us in the real world, the natural world, where we belong.

The Purple Emperor season challenges us to delve into – and, crucially, become part of – a more wondrous reality, which exists at the fulcrum of nature, at the very zenith of the natural year. The relief is staggering. We even sleep properly at night, and not simply because Emperoring involves long days spent standing up, loitering with intent, forever concentrating. It is not just physically exhausting, especially day after day, but profoundly fulfilling. The remainder of the year scarcely matters, being at best a build-up, or a come-down.

This butterfly reaches the inner core of the human psyche, which other elements of our wildlife experience struggle to reach. Not for nothing did the ancient Greeks believe that the human soul departs from the body, on death, on the wings of a butterfly; not for nothing was Psyche, their goddess of the human soul, named after their word for a butterfly, psyche – hence psychology, psychotherapy, psychometric profiling, and so on. Yes, butterflies are intrinsically linked to the human soul, and the Purple Emperor is quite simply our ultimate butterfly, and should be recognised as England's national butterfly. Other butterflies establish colonies or metapopulations (clusters of loosely connected colonies functioning at landscape level): the Purple Emperor establishes an Empire, which aspires to encompass the human mind. So, this butterfly is a metaphor for something deeply and fundamentally profound, not just about humanity or nature, but about existence itself. Oh, to understand metaphor! – and to have it taught in schools alongside metaphysics.

With these feelings of immense relief comes a deep, ineffable joy, a contentment as real as July's blue-green Oaks, as perfect as the Foxglove bloom and as heady as Honeysuckle fragrance. It is the lifeblood of my soul, and I am not alone: these feelings are shared by the growing number of us Emperorphiles who have become known as the People of Purple Persuasion, and who seek, through Emperoring, to become part of nature herself. It is not the only way in: there are others, somewhere. Find yours – by following the song-dream. We, even more than the land, desperately need rewilding. The Purple Emperor season shows us the way.

Knepp Wildland, 5th July.

Surely I dreamt to-day, or did I see
The winged Psyche with awaken'd eyes?
I wander'd in a forest thoughtlessly

John Keats - *Ode to Psyche*

Evening Flight

And I shall ask at the day's end once more
What beauty is, and what I can have meant
By happiness?

Edward Thomas - *The Glory*

All aspects of the day had been perfect, for it was the high point of summer. The heat of the afternoon had grown too hot for many of the Wildland insects, who had retreated into the deep shade, into the places frequented by shadows numberless. It had grown too hot also for the Tamworth pigs that roam freely within the 470-hectare (1,128 acre) Wildland block, rootling hither and thither, carefree as the cotton-wool clouds above. They, too, had retreated – to the shadiest of hollows where some vestige of oozing mud remained, and collectively had fallen into soporific slumber: the greater the heat, the deeper their stupor.

But as the temperature gradually subsided, and the power of the sun's rays weakened, the insects and pigs awoke – piglets first, then sows. Hoverflies recommenced their hoverings and gnats and midges began to dance. Shafts of sunlight, relieved of menacing heat, became alive with hosts of floating insects: in places so dense that they seemed like wisps of curling smoke that were almost elasticated, even ghost-like, as they rose, fell, gyrated, and levitated within the most solemn of dances. In places they seemed like particles of living dust: the fundamental particles of the food web. And they offered their music too – a soft but throbbing hum and the faintest of whines, which made the Wildland air gently pulsate. The bulk of these small beings were a type of winter gnat which flies in high summer, but many kinds were there, and not just those who live but for a day, the ephemera.

And where the early evening air hung still, along the westward-facing Oak edges, butterflies that seemed like grey sprites of this frenzied vibrancy, began to circle dance. These were Purple Hairstreaks, perhaps the most numerous butterfly in the Wildland: dark iridescent purple on top, but with smoky-grey undersides – which is all that we below see of this enchanted, and enchanting, butterfly. They breed on the numerous Oaks that line the old field boundaries; but after a flurry or two of activity as the morning warms up, they spend the day in a near-comatose state up in the Oak sprays, waking after 5pm to commence their evening flight, and to conduct their all-important courtship and mating. They are most active between 5.30 and 7.30pm, but may stay active till way after 9pm on heat-wave days in early July, when the temperature has soared towards the old-fashioned 'nineties' (30° Celsius in modern language).

This evening they were wondrously active, for the day's heat had stimulated the year's main pulse of emergence, and numerous virgin females were in need of mating; and for some reason or other this was clearly a good year for this small and under-recorded butterfly; after a run of lean years, which had brought meagre emergences. Densely-foliaged Oaks revealed half a dozen, or more, in a vista composed mainly of pairs: frenetically circling around each other, spiralling almost out of control, out and away from their tree, like dancing Sufis, whirling dervishes, before returning, still circling, to their chosen tree heft. As many as two dozen individuals were active over favoured trees, though sparsely-leaved Oaks, with yellowing leaves, at best attracted only single stray or over-spilling individuals. In places along the old green lane they also flew around sunlit Ash trees, where they obtain sustenance from the sticky secretion offered by the developing buds of next spring.

Acting perhaps like winged flowers, they circle-dance the living air, in celebration of midsummer's days, seeking a life beyond an evening of perfection. The song-dream had entered every living being this evening.

Then drift, my spirit, drift, in sublime timelessness, suggesting all and everything to dancers in the mysteries of life, that drift into a state of undying, before being seen and felt no more. But ask not what orchestrates this dance, for that much is hidden.

Knepp Wildland, 7th July.

Thistles spur the meadow in July
unpack their blazoned heads.
The oaks are crowned
with honeydew and the air-
dance of this colony,
indigoed, extraordinary.

Jean Atkin - *Purple Hairstreak*

The Great Stank

There are times and moods in which it is revealed to us, or to a few among us, that we are a survival of the past, a dying remnant of a vanished people...

WH Hudson - *Hampshire Days*

The forest closed around me. The outside world instantly cut out, irrelevant, deceased – Lyndhurst and its perennial traffic jams; and vehicles rushing towards the Lymington ferry termi-nal as fast as a myopic road, built for slower, dustier times, would carry them. All I had done was venture through a wooden forest gate; but it felt like entering another dimension. Welcome to Pondhead Inclosure, the last of the New Forest woodlands to retain its true character, and the only one of the forest woods which the Victorian and Edwardian butterfly, moth, and beetle collectors would still recognise: a place of gigantean lichen-soaked Oaks and high Beeches towering above narrow, Bramble-lined rides filled by long shafts of sylvan sunlight. The other inclosures, all created primarily to supply the navy with Oak timber, were destroyed by the 20th-century forestry revolution and have been replaced by dark, brooding, and labyrinthine conifer plantations – England's Black Forest – or staring acres of straight-limbed Beech, devoid of underwood and any vestige of spirit.

I hadn't meant to come here. I was going to another Hamp-shire forest, but got hijacked – The Forest was calling me – she doesn't take no for an answer and I love her too much to refuse her. Places do that: they suddenly call you in, especially those seeped-deep in ecological and human history, like ancient woods, and especially places well-inked-in on the spiritual landscape in which our lives are enacted. We love such places with an almost ineffable love, and I have an idea that that love is somehow reciprocated, though don't ask why or expect rationality here: love, like faith, is irrational. But I am convinced that our relationship with special places is two-way. The urge is to venture deeper – into the secret pulsating heart of the forest – for we were once forest people, until we unwittingly cleared it.

Beechen Lane, the old drove road that separates Pondhead from its neighbour, Park Ground Inclosure, is edged by a scatter of veteran Oaks, many of them with dark smear-stains, a metre or two long, running down their lower trunks. These are the vestiges of treacle patches, painted on by moth collectors, from late Victorian times onwards into the 1960s. Each collector had his own secret recipe, almost invariably involving treacle, sugar, and alcohol; often rum, plus some secret idiosyncratic ingredient. They would stand guard over the trees they'd bagsied, and net or box drunken moths. The rivalry, skulduggery, and social snobbery that these collectors practised was monumental, but has long sunk into the nether regions of entomological history. WH Hudson loathed them all, for nearly all of them were takers, not appreciators, let alone givers-back. Yet I yearn for the experiences they must have had, for the richness of the forest of yesteryear.

Every now and then someone splashes something new on to one or two of these inveterate treacle patches, in pursuit of the giant Crimson Underwing moths that the collectors so deeply prized: species with mnemonic names like *Catocala promissa* and *Catocala sponsa*. However, today's burgeoning number of moth enthusiasts catch their moths in sophisticated light traps, and the art of sugaring, as it was known, has largely been lost. Gone too is the knowledge many of the old entomologists would have had that *Catocala* is derived from the Greek words for Beauty and Underneath. Long gone is the gentle blizzard of moths the collectors of old would have encountered on their nocturnal perambulations under star-filled skies, their way lit by myriad Glow-worms, their bodies battered by airborne Stag Beetles, their lanterns picking out the eyes of distant Deer, Foxes, and Rabbits.

Pondhead Inclosure has witnessed all this, and much more. Visiting in mid-July, when the collecting season would have been at its peak, and all the new red-bricked board and lodging houses in Brockenhurst and Lyndhurst taken by earnest collectors, I feel not of today's age, but of a halcyon bygone time. I am walking amongst ghosts, the ghosts of my dispositional ancestors: the butterfly admirers of yesteryear, clad in frock coats or Norfolk jackets. I feel that they sense me, and that they view me as being one of them; even though I could not kill a butterfly to save my life, let alone pin one under the crown of my tall hat, as some Victorians did. Despite my abhorrence, I feel closer to those of them that were competent naturalists than I do to my genetic ancestors: Cornish miners and Methodist ministers, Strict or Wesleyan, and their dour and doughty wives.

This morning heat was intensifying as atmospheric pressure rose after a thundery night, and the forest was steaming. Everywhere pillars of vapour were rising; some contorting into semi-human shapes, before dissolving. The birds, though ever present, were largely silent, flitting, skulking and flying from darkness out into blinding light, and back again. A Wood Warbler trilled briefly, and belatedly, up in a great Beech, for it was far too late in the season for him to be calling. He must have failed to breed. A family of young Redstarts, in muted colours, were learning to feed amongst the lofty branches of a centuries-old Beech. From above the high-flung canopy a passing Buzzard mewed, and was gone, unseen.

Everywhere there were insects, buzzing, humming, tapping and vibrating – such that the forest itself pulsated. A gentle rain of caterpillar frass dropped down from above, Dor Beetles bumbled along the ride edges. The Tree Slugs were out too: crawling round to the shadier side of the tree trunks, where the moss was damp. Wood Ants foraged on every inch of the forest floor.

A posse of grey flies gathered above my head, and followed my every movement along the bare clay rides. These were the Head Fly *Hydrotaea irritans*. Aptly named, they do not bite or suck blood, though they will drink it from open wounds. They cause summer mastitis in cattle and worse in sheep; and are vectors for the infamous New Forest Eye, which is caused by a bacterium and can lead to animal blindness. With humans, these flies seek primarily to imbibe sweat, but succeed mainly in causing severe irritation and mind-numbing distraction. They could be used in torture, suitable for ancient Greek antiheroes like Sisyphus or Tantalus. They used to occur in such plague proportions in the forest that collectors would smoke pipes continuously to drive them off, or smear their skin with tobacco tar. Some even wore beekeeping veils. WH Hudson knew them too, writing in *Hampshire Days*: 'The oak woods were now full of a loud continuous hum… an unbroken sound composed of ten thousand thousand small individual sounds co-joined into one, but diffused and flowing like water – the incredible number and variety of blood-sucking flies.'

Forest legend has it that their populations collapsed after the great drought summers of 1975 and 1976, which begs the question of just how numerous they were before then? There are, of course, impediments to paradise on earth, for this is a fallen world.

In late morning the horseflies became active blood-seekers: small grey clegs, whose bite can cause hurtful swellings, silent hunters of exposed human flesh. As it warmed up, a Dark Giant Horsefly *Tabanus sudeticus* female hummed past, the size of a Hornet queen and intent on sanguine malevolence. The forest Fallow Deer would suffer today. And the Hornets too were up: hunting hoverflies, bees, and butterflies amongst the flowering Brambles, and zapping in-and-out of their nest in a Woodpecker hole 20 foot up in a gnarled corner Oak. They're not interested in us, unless we stand in the flight line close to their nest. Another giant was hunting jewelled Dolichopodid flies along the ride-side ditch: the Gold-ringed Dragonfly, which breeds in the forest's gravel-bottomed streams.

Speckled Wood butterflies were hatching along a narrow ride studded with skylight pools – a butterfly of dark and light, whose markings are perfectly in tune with the dappled shade they inhabit. A pair of males were squabbling over the possession of a sun spot, circling continuously round each other until one, the weaker, gave up and moved on, leaving the victor to claim his shaft of woodland light, and migrate with it, as the sun moved the day onwards.

There are giants in the butterfly kingdom here: a huge brazenly-joyous creature that comes in three colour forms and which we know as the Silver-washed Fritillary, but which the old collectors knew as *paphia*, for they used only the Latin and Greek names, believing that these names imply greater reverence. In days of yore this butterfly abounded in many of the New Forest woods to a degree beyond our furthest imagination. There are descriptions of them descending from the trees, after a shower, in such numbers that they darkened the sky and seemed like a heavy fall of autumn leaves. There are also reports of picnickers becoming covered with their spiny bronze larvae. The collectors greatly prized the dark iridescent-green colour form of the female, known as aberration or variation *Valezina*, which in those days was regarded as a specialist of the New Forest. *Valezina* is a creature of the shadier rides, Speckled Wood country. The commoner, normal-coloured, fulvous-orange females, and the bright saffron males are creatures of the hot sun. All are lovers of Bramble flowers, indulging in a mass mid-morning breakfast along the Bramble-lined rides, before dispersing to explore the secret parts of the forest, or dancing off as courting pairs to mate for an hour or two high in the canopy.

On this late morning, in a modest butterfly year, *paphia* was out and about only in ones and twos, hither and thither, with at best only three or four on the sunniest Bramble patches in fullest flower. Some had already torn their wings amongst the thorns. At the western end of Pondhead's main ride, on a low hilltop, close to Beechen Lane, stands some of the largest Bramble patches in the whole forest. Here one can wait and watch, for each of the Silver-washed Fritillaries in the region of Parkground and Pondhead is duty bound to visit this clearing, a place of calling. Others come too – WH Hudson's favourite butterfly, the darkling White Admiral, which the old collectors knew as *sibylla: Sibylla* of the Brambles, *Sibylla* of the woodland rides; and which skims close to the foliage with impressive agility and entrancing grace, together with all the humble Ringlets and Meadow Browns, and the forest Honey Bees, small and dark, and the myriad hoverflies, and the hunting haunting Hornet.

Then from amongst the leaf litter under the Beech edge, where a few strands of Common Cow-wheat twist and turn their tubular lemon-yellow flowers, came the soft, contented, purring of the Wood Cricket. It wafts within, rather than on, the air. It does not sound British, let alone English, but sounds more like something out of a western movie – in the nightfall moment before the Apache Indians attack. Yet it is the sound of the New Forest in high summer: in the warmth of the day and in the shadows of the sultry forest night. It is the song of the forest earth.

Wandering eastwards along the main ride, which is fringed with Bracken and tinged with golden light, one can veer off to the right – down the vestiges of an ancient ride now all but closed over and frequented only by the forest deer. There is a hilltop, loosely crowned with Oaks, where one can scan for Purple Emperors, for this was a known haunt of *iris* a century or so ago. The experienced collectors knew, for they worked these woods assiduously, and their field-craft skills were exemplar. A Stock Dove was calling: more a series of coughs and whoops than anything approximating to a song; yet a paean of the veteran trees.

Descending downslope, beneath Oaks as aloof as if we were ants, one comes to where the ground levels out and becomes boggy, somewhere towards the hidden wood edge. Here, in the great drought summer of 1893, a summer seemingly more extreme than even the long hot summer of 1976, a mighty vision was seen. In those days there was a shallow peaty mere here, fringed by Sallows, called The Stank or The Great Stank – but it is long gone now, having become a large tree-filled depression. Back in the mid-July of 1893, a couple of young butterfly collectors found their way here, in hope of bathing water; but they found the pond dried-up and cracked, and slotted with the hoof prints of deer and pigs, which had wandered there to drink or wallow. And instead, their eyes were astounded by innumerable butterflies that had descended to imbibe the last vestiges of moisture. The bulk were Silver-washed Fritillaries and White Admirals, feasting with closed wings in searing heat – the silvered-green undersides of the fritillary and the silver-and-tawny of the admiral, *paphia* and *sibylla* – in profusion. There too was *adippe*, the golden High Brown Fritillary which breeds on violets within the Bracken beds, and *polychloros,* the elusive Large Tortoiseshell – both these now long extinct in the New Forest, and elsewhere. Even the god *iris* had descended from on high to drink: a pair close-by, feeding with closed wings, but flicking at pestering flies. And there were others too: humble creatures like the Green-veined White, the azure Holly Blue and the golden Small Skipper; and beside them a host of lesser insects – bees, beetles and flies, forest creatures all, revelling within the moment of being within a great summer's thirst.

This is what we have lost, even from a place as large, intact and legally preserved as the New Forest. We need to rebuild the Forest, to renew ourselves as forest people, to find our way back and become its children again. That is where the song-dream leads.

Pondhead Inclosure, New Forest, 16th July.

The tall forest towers:
Its cloudy foliage lowers
Ahead, shelf above shelf;
Its silence I hear and obey
That I may lose my way
And myself.

Edward Thomas - *Lights Out*

Great Uncle Harry

I thought to follow him myself. But the next day I was still in that grey land, looking at it from a railway train.

Edward Thomas - *The Heart of England*

Mindless summer cloud greeted me at Platform Zero, at Stockport station – the sort of cloud that obfuscates an anticyclonic day in late July, before dissipating in the evening. Unsullied, Swifts speckled far overhead, and a lone Large White butterfly flopped on to a Buddleia panicle, protruding through the platform fence. Below this lone flower lay a half-drunk bottle of Gordon's Dry Gin.

Fireweed jungles told of embankment fires in days gone by, when steam trains ran the Stockport-to-Buxton line, sparking trackside grass fires. Reddening Rowan berries and whitening buds of Russian Vine whispered of autumn. My train, noisy in the extreme and overheated to the point of absurdity, threatened to expire at every station, before spluttering on – just – past Bramble clumps that paused between the ending of blossoming time and the greening up of fruits.

Middlewood Station offered canary-yellow Evening Primroses and, on the bank past it, the soapy pinks of Himalayan Balsam, with a Cabbage White every 300 metres. Disley, the station for the National Trust's Lyme Park, had a pocket of station flowers and an unprepossessing tunnel that still reeked of steam-train smoke. Behind a new housing estate, devoid of chimneys, the inclines and flattened summit of the Peak District beckoned. My Great Uncle Harry had died out there, alone, at New Year in 1922, aged 50. A bachelor with an interest in poetry, fluent in French, Italian, and for some reason Flemish, he had gone off walking by himself, as was his strongest want. His body was found after a week-long search, on the Edale side of Kinder Scout, in a rock-strewn gully known as The Downfall: a landscape akin to that of Mordor. One obituary included: 'He asked for, and must have got, his fill of the oldest and most elating of combative joys.' A studio portrait of him went on sale as a postcard, and raised so much money that a plaque was erected in his memory in Hay-field churchyard. His death exacerbated matters leading up to the Kinder mass trespass of 1932, for the landowner wrote tactlessly to the *Manchester Guardian* lamenting Great Uncle Harry's presence on the moor, and emphasising: 'The public have no right of way on Kinder, over Kinder, or through The Downfall', and added that the extensive search had 'depreci-ated the sporting value of the property.' I decided to resist Kinder's lure, and remain on the train. Great Uncle Harry could have it: he'd earned it.

After glimpses of the muddied waters of the Peak Forest Canal, Whaley Bridge Station restored one's faith in England with proper station flowerbeds, of *Alchemilla mollis*. But the whole line cried out for the return of steam trains. It didn't want to be part of 21st-century England, it wanted out, of modernity: and more Cabbage Whites, more dead tractors nettling away in fields of Yorkshire-fog grass, and the return of milch cows. It wanted time, not just to stand still at the advent of August, but to run backwards into a more glorious era – one Great

Uncle Harry would recognise – of Buttercup fields and abundant Meadow Brown butterflies, farmland birds, and the haunting cries of Curlew.

The situation worsened at Chapel-en-le-Frith: the station for Eccles Pike and Combs Reservoir, for its hanging baskets held plastic flowers. Insulted, no one got off. The train coughed on, disturbing a pocket of sooty Ringlet butterflies amongst a stand of Tall Fescue grass, before vanishing into another soot-stained tunnel to crawl through cuttings of Dropwort, Marsh Valerian and yet-to-blossom Fireweed.

Dove Holes Station boasted a Mesolithic powerline, a scrap-yard full of rust, and a vast desert of hill-slope pasture rendered soulless and devoid of wildlife by far too many sheep – one could die out there, of despair, or tedium. The sun flickered wanly through an unnecessary pall of cloud, briefly and retro-spectively, as Buxton beckoned. The train expired there a mere three minutes later, perhaps for its final time.

England's highest town, Buxton, exists within its own enclave in time. It is still proudly Victorian. Great Uncle Harry would recognise it, happily. You can hear its river and Jackdaws, rather than traffic. The town sports Victorian flowers: Astilbes, Bistorts, Hostas, and red-and-white floribunda roses. Below the Opera House nestles the Pavilion Gardens, a place apart – with Daisy chains, water gardens, Petunia baskets hanging from Narnian lampposts, a miniature railway that runs on summer bank holidays, a summer festival bandstand, and Lime trees standing still – and everywhere, Jackdaws.

Someone needed to take charge here: to time-capsule the town forever in late July, in the late Victoria era. The Jackdaws had seized the day, and had taken charge. Although the clock read but two-thirty it was time for Earl Grey tea and fruit scones, with jam and cream. Buxton and its Jackdaws both expected and orchestrated it. I had been re-reading TS Eliot's *Burnt Norton*: one of the *Four Quartets* meditating on the poetic philosophy of time, but I could just have easily have reached the same conclusion on the meaning of time by watching the Buxton Jackdaws.

This chapter is dedicated to the memory of Henry 'Harry' Fowler Martin (9th August 1871 to circa 1st January 1922): gentleman, moorland wanderer.

Buxton, Derbyshire, 23rd July.

There's a kind of line, of light, a thought line, which cuts through false histories and comes towards us from the devastated zones. When it gets to the British hills it twirls round a stem, if it can find one.

Peter Riley - *The Ascent of Kinder Scout*

Ashford Hangers

Perhaps
I may love other hills yet more
Than this: the future and the maps
Hide something I was waiting for.

One thing I know, that love with chance
And use and time and necessity
Will grow, and louder the heart's dance
At parting than at meeting be.

Edward Thomas - *When First I Came Here*
(Shoulder of Mutton, 1916)

All spring and summer, in the false sense of hurry we live largely within, I had passed through Edward Thomas's heartland – the hanging woods above Petersfield: between home (whatever that is) and Knepp Wildland (which had become my natural home); to and from, betwixt and between, always in a hurry to be elsewhere soon; when here was now, just here and now, and now was always here; and the place was calling me.

Then, one late July afternoon, whilst heading west towards the sunset of a summertime, Ashford Hangers finally arrested me. It said: "Stop. Spend some time with me now. I have something to say – not necessarily to you, but perhaps through you." So I turned right at the top of Stoner Hill, past the u-bends, into Cockshott Lane, past the site of Edward Thomas's Arts and Crafts studio: to bump and scratch along Old Litten Lane until it petered out. Within seconds of exiting my dust-covered car I'd long forgotten that I even had a vehicle – it no longer mattered.

Time ago, I had lived nearby, for twenty years in fact; but that period was now itself some twenty years past. Back-along, I had wandered every path on and around Ashford Hangers: every track, every Deer or Badger run, or so I thought. I knew it all, or so I thought. I had surveyed the hangers' insect fauna, finding rare species associated with decaying wood, anthills, calcareous springs, seepage marsh, and more. I had spent weeks searching for rarities, for today's naturalists are obsessed with rare species and habitats, though those are only two attributes of specialness of Place. And although I had only scratched the surface, those surveys had helped make Ashford Hangers into a National Nature Reserve. All that had made me feel better, but now it felt that that designation only belittled the place: it is a poetic landscape of at least national significance.

All the while, the spirit of the place had insidiously seeped into me, perhaps as it had done eight decades earlier to Thomas. At times I had searched for Thomas's spirit; but he was not there, he had only passed through, after all, and the place had used him: it had spoken through him, and moved him on, though keeping some vestige, some ghosting, of him as part of itself.

That early evening, I sat down amongst the flower-studded grasses of the steep east-facing slope of Shoulder of Mutton, with the hazing Weald and the western South Downs stretching out before a sunset that grew and flamed somewhere behind me, in a sky lost behind the treeline. The dog walkers had all gone home. Perhaps they had been summarily dismissed. It felt that I alone had been invited. We must all have that feeling, from time-to-special-time. I hope so.

To the right, the downland slope was fringed by a line of Yews, viridian-green in colour, dusty, morose and utterly silent. To the left, their crowns still in sunlight, ran a line of tall Beech trees: leaves gilding towards autumn. Everywhere, the flowers of advancing summer on the chalk downs – Clustered Bell-flower, Common Eyebright, Common Knapweed, Felwort, Field Scabious, and Common Ragwort – amongst the dead-heads of spring's False Brome Grass; and everywhere too, the purpling and bronzing leaves of low-growing Dogwood. A pocket of Common Blue butterflies was roosting, head down amongst the grasses, loosely and diffusely. Dark-bush Crickets were calling, but these I could sense more than hear as advancing age is depriving me of the ability to hear the songs of our crickets and grasshoppers (this constitutes the elimination of a natural right).

A poem was born then and there. I did not write it, and I may even have watered it down, or spoilt it: the place wrote it, through me, as a tribute to one of its lost sons, who found his heartland there –

Steep

Only in time, for there is always time, between
The showers that skim the leaf-unfurling green,
As rainbow-hailed the land slow dries,
And sends to dance again the hoverflies,

Or stills the stormcock bird, who rings his song
In the temple of the stilled winds, and belongs
Just there - where the Weald stretches out its clay,
To greet a vagrant swallow, arrived today,
As if to say, you have returned, from France,
Perhaps, or some place safe from distance
To your heartland, where the woods hang down
Their heads to greet one who wears a poet's crown,
Who drifts, as sallow seed, somewhere above
The yew tree slopes that bore your steeping love.

In Lutcombe, where wood forget-me-nots
Hide in shade as spring's leafing sudden stops;
Up Wheatham Hill, where the beech cathedral
Rallies a faithless faithful with a cuckoo call;
Down Honeycritch Lane, as youth out courting
Lingers, while the snickering of bats begins;
In Happersnapper, where autumn vapours hang
In misted drifts of sodden leaf-mould rain;
On Shoulder of Mutton's thyme-stained down
Where you would rest on June days, bygone,
So long bygone, but choose one moment –
Choose any moment to begin, again, re-sent
As one restored to life by the poetry of love,
Bidden by the calling of some hidden turtle dove.

For here, between the pigeon's breasting dive
And spring's unravelling of dandelion lives,
Or the flashing of iridescent time on insect wings
High in the oaks, as July wanes its evenings;
Just here, within the moments of memory,
We can choose our moments, and simply be,
Even as you, as distant blue smoke rising
From a Wealden charcoal fire, an accident of time,
Or as the striking moment, when sunlight muddles
From a mirror masquerading as a puddle;

Or in footprints of children's laughter, imprinted
In the mud, the dust, of time, indented.
Remember! For it is written so, and right,
Within the lucidity of light and twilight.

For you found in Nature here, the words of life,
Gave them shape, released them, into the narrative
Of memory, then marched off east-away,
To the chalky mud of khaki France, away
From a formless war within your mind,
Beyond summer's haze, in tears of autumn rain,
Or in pursuit of spring, as the seasons lied
Within a soul confused by mere humanity,
That pondered what is meant by God, or love,
When all around some spirit subtly wove
The wonder that you saw, breathed and conveyed,
And offered truth, unbroken and unmade,
And formed perfect into eternal words
The messages you thought were yours.

So you acted out your ministry, perhaps unwittingly,
Not knowing it for what it was, and ever yet may be;
You took Nature's meaning in your hands,
Blessed it into words, loosed it upon this land;
And you said one morning, on waking from a dream:
"Somehow, some day, I shall be here again".

Matthew Oates - *Steep*

Shoulder of Mutton, Hampshire, 26th July.

Witherslack Ascending

Prophets of Nature, we to them will speak
A lasting inspiration, sanctified
By reason and by truth; what we have loved
Others will love; and we may teach them how;

William Wordsworth - *The Prelude* Book XIII

What do you do when the party's over? Find another party, even though it can scarcely hope to match the Bacchanalian splendour of the one now spent. Nothing rivals the Emperor season, which means that nothing rivals the pathos of its ending.

The M6 was full of caravans heading up to the Lake District, or the great beyond. Its service stations were in desperate need of obliteration. But from the summit of the long rise up past Lancaster University, the grey limestone hills of Morecambe Bay shimmered and beckoned. My heart leaps for joy on breasting that rise, for the grey hills greet me, in a prodigal-son moment.

Arnside was full of active retired people, fit and amply-pensioned, and its skies full of Swifts. The active retireds were drinking, recollecting the day's perambulations and discussing tomorrow's. The Swifts were screaming and reeling in the early evening air, racing each other from here to Pigling Bland's hilly land and back again (or wherever it is they go to in Beatrix

Potter country across the River Kent's silvery estuary). It was all too apparent that soon they would be off altogether, southwards and beyond, without saying farewell, leaving the black painted eaves and gables of Arnside's Victorian houses bereft, and the evening skies vacant.

The state of Arnside's gardens told of a summer that had misfired, and of the imminence of August: the Sycamore trees were darkened by Tar Spot Fungus and hinted that summer had little power left, whilst flowerbeds of Japanese Anemone and Fuchsias spoke almost of September, and were devoid of bees and butterflies; the Sweet Pea canes were wind-bent and window box Petunias were flowering only sparsely; the grass was still green, having not been exposed to any drought stress; and everywhere was verdant with moss. Towards the estuary, and seawards over Far Arnside and Heathwaite, the Sycamore leaves had been scorched and withered by salt burn, blown up on summer storms. All this told of an inclement summer up here. Worse, ripening Rowan berries and the silence of the House Sparrow families heralded, or at least anticipated, autumn. The north-westerly breeze was dying, to usher in a cool night under a sky that promised the Milky Way and Perseid meteor showers.

But the morning was bright and fair, though it was obvious that there would be no heat in the day's sun. It would cloud up inland, over the distant Dales, and over Lakeland, and the Pennine fells to the north-east, but around the coast it would remain clear. Across the water, beyond a slumbering Grange-over-Sands, Humphrey Head was standing out for sunshine, and twinning itself with the more famous Worm's Head off the Gower coast, over in the Land of Song. North-eastwards, the southerly crags of haughty Whitbarrow were illuminated, golden and lionised, in morning light.

The lower slopes of the long ascent up to the Knott from the town are known as Redhills. Years ago they hosted a golf course but are now steadily reverting to mainstream nature, under the gentle care of a few Highland cattle who reside there all year round. The Bracken beds are expanding and infilling, providing a habitat for fritillary butterflies and breeding Willow Warblers. Antler moths were clustered on Common Ragwort flower heads. High up, on a scree spur, stands a loose clump of four isolated Scots Pines. These pines define the spirit of the place, yet no one visits them – they are not 'calling' trees; they do not call to us, but sing with the Aeolian winds instead.

The way on to the hill top from Redhills is through a narrow gate of Hazel spars in a high-set boulder stone wall. Passing through it feels like crossing into another world. The Knott, that morning, felt wind-rocked, a place of Yews slanting lee-ward: many of them topiarised into curious shapes by browsing Deer, prostrate Junipers, Heather pockets, beds of limestone scree and rock, and broad corridors of Blue-moor Grass. Southern Wood Ants were writhing about on their gigantic mounds of arboreal bric-a-brac, which can be likened to Dalek supreme-command stations. I am determined to find the Scarce Seven-spot Ladybird *Coccinella magnifica* up here. An aphid predator, it is strongly associated with *Formica* wood ant nests, but is not known to occur nearer to Morecambe Bay than Cheshire; but if it does occur anywhere in the north it will be in Morecambe Bay, where north and south meet ecologically. I will keep looking, for nature pushes limits, including its own, and naturalists should follow suit. But at least the Dark-red

Helleborine (or Royal Helleborine) was still in flower, in places where scrub meets scree and where the Roe Deer had not eaten off its flower spikes.

A meagre-few Scotch Argus butterflies were bumbling around, low over the grasses on the north-facing slope, clad in dark velvet moleskin wings with crimson spots. It is remarkably large for a member of the Brown family: a cousin of the humble and ubiquitous Meadow Brown, which it rather dwarves and outscores. It is also highly wary of people, much to the chagrin of the butterfly photographers who travel up to Arnside Knott from southern England to photograph this northern species at its most southerly haunt. Each Thistle clump was tenanted by at least one photographer, clicking away and shouting out inanities to his companion, like: "Nah! These are too tatty." The much vilified butterfly collectors of yesteryear had the same habits and spoke the same language. Insulted, the butterflies dispersed over the wall. I decided to follow their example and retreat to somewhere where one can be alone with nature. Quite often, one can find oneself in the wrong place, or in the right place at the wrong time – especially when solitude calls, and it does call, mightily.

The northern side of the Kent estuary is another world, at least until you reach the retirement centre of Grange-over-Sands. Tucked away between the limestone massif of Whitbarrow and the long winding hill of Newton Fell, on acidic Silurian Slate, is the rather disparate village and civil parish of Witherslack, crowned by a low wooded hill, Yewbarrow. You can lose yourself, and everything else, on Yewbarrow. Time itself instantly dissipates there, and summer lingers long – longer than across the river, around Arnside and Silverdale. Someday this hill will dictate a decent poem to one of its sporadic visitors, who seem to be there *by invitation only*.

A narrow path ascends the steeply-wooded slope from opposite the Old Vicarage – now an hotel offering Tranquillity (with a capital T), and close to the 17th-century St Paul's church (which is kept open). You lose yourself in somnolent Yew woods, then zig-zag uphill, along a multifarious path system which ensures that you never go the same way twice. If, breathless, near the summit, you take the correct fork, you come out on top of the exposed limestone scar and can look almost directly down onto the church tower, which almost always flies the Cross of St George flag, and on afternoon tea being served in the Old Vicarage garden. Much of the best of northern England lies spread out before you: Arnside Knott, somewhat apologetic across the estuary; the vast raised bog of Meathop Moss, with its waving cotton grass heads and wild Cranberry beds; Halecat Nursery and surrounding coppiced woods; and innumerable low hills, winding northwards and upwards, towards Coniston. Somewhere up there, in the churchyard at Rusland, that champion of childhood freedom in nature, Arthur Ransome, lies buried.

The scar top itself is carpeted in Common Rockrose, which supports a colony of miniscule Northern Brown Argus butter-flies. Normally an insect of June and early July, they linger into August up here, where the flight season is later. Long ago, I think in the early 1980s, I watched a Hummingbird Hawkmoth flying along the crag face: a female dashing about and deposit-ing her soft green eggs on Ladies Bedstraw leaves, growing in crevices amongst the rocks. Yewbarrow ensures that moth is remembered on each of my sporadic visits.

Wandering on, and inwards, one pushes through a thick band of old coppiced woodland, Ash and Hazel in the most, before arriving suddenly into a hilltop expanse of rough Blue-moor Grass grassland, heavy with ancient Yew trees. Welcome to the secret world of Yewbarrow – a place of withdrawal from the maelstrom world. I first ventured here back in the long hot summer of 1976, at the end of that flaming June. It was lonely and neglected then, and still is, for sheep grazing ended here decades ago. Yet it has changed, monumentally: dead grasses have built up to suppress the Violets on which thriving populations of fritillary butterflies bred, the scrub edge has moved ever inwards, and the Yews have grown massively. It needs grazing, lightly, by cattle, during the summer months when the beasts are unlikely to poison themselves by eating too much Yew.

Moving north, just before the old grey wall that divides Yewbarrow from its larger sister, Strickland Hill, is the vestige of a grassland glade – fringed by Bracken beds and Hazel scrub, and populated by ant hills. Here, over the years, I have studied fritillary butterflies, found their caterpillars, and watched flycatchers feeding their young – oblivious to what was going on in the world outside. One day, some of my ashes will need to be scattered here. There are places, many of them mentioned in these chapters, which will require a scattering in due course: I know that one of my daughters will do this.

Strickland Hill used to be grazed heavily each summer by hill sheep and was consequently rather void of wild flowers, other than plants like Common Bird's-foot-trefoil and Wild Thyme which belly-crawled below sheep-bite level, and was inhabited by relatively few insects: Common Blue, Northern Brown

Argus, and some Grayling colonies amongst the pockets of scree. Then the sheep regime ended, and the flock was replaced by a modest number of traditional-breed cattle, who wander around in summertime, at ease with time and place. Their actions have transformed flora and fauna here, so that the 'wildlife interest' has actually migrated from degenerate Yewbarrow next door, to the revived open limestone pastures of Strickland Hill.

To the east, across the tops of dense Oaks growing in a shallow valley, the massive two-mile long flank of Whitbarrow Scar's crag and scree system is illuminated, cathedral-like, in evening sunlight. Whitbarrow seems too vast to be English; it should be Pyrenean or Dolomite perhaps, yet it stands there, tall and proud, a place where nature is still in charge of its own kingdom.

To the north the limestone hills are replaced by other low hills, on acidic rocks, with Bracken, Heathers, sheep-walk and silage fields. These new hills undulate away into Lakeland, and an unexplored beyond. In time they will call me in.

Witherslack, Cumbria, 2nd August.

Wordsworth's Grave

The world is too much with us; late and soon,
Getting and spending, we lay waste our powers;—
Little we see in Nature that is ours;
We have given our hearts away...

William Wordsworth - *The World is Too Much With Us*

It is hard to determine why so many people visit the Wordsworth family graves in the ancient graveyard surrounding St Oswald's church in Grasmere. As a poet, William Wordsworth has relatively few genuine disciples today, far fewer than many of the more recent poets, such as Dylan Thomas and Philip Larkin, and, from amongst his Romantic peers, fewer than John Clare and John Keats, whose bright star is now very much in the ascendency. Few of the people who stand and stare at the graves can recite more than a couple of lines of Wordsworth, or name the title of more than one of his perhaps-too-many poems. They stand bemused, not knowing who all those Wordsworths were, knowing only William and Dorothy, and perhaps Mary. It's not even as if they've been put off Wordsworth by being forced to spend an entire term studying the Lucy poems in their teens, as older generations were (those poems are not teenage material, they are insufferable at that age and should not be inflicted on young people).

It's not simply that Grasmere, where Wordsworth lived from 1799 until his death in 1850, has been skilfully placed on the Lake District tourist map or that his main places of residency there, Dove Cottage and Rydal Mount, are being cleverly marketed, so that people *do* Wordsworth in the same manner in which they *do* Beatrix Potter, Windermere, Ullswater, and the Langdale Valley. The irony is that these places depend on mass tourism for their survival as business communities: yet Wordsworth raged against the development of mass tourism in his beloved Lakeland, and the coming of the railways – despite making most of his of his modest fortune from his *Guide to the Lakes*, first published in 1810. Perhaps visitors to Grasmere should be asked: what are you seeking here?

The truth is that Wordsworth's name remains revered because his values, and his way of looking at the world, somehow reach out and touch a special chord within us, even today. We are aware of the depth of his relationship with Nature (he almost invariably capitalises the word). His name stands out as a symbol against all that has gone wrong between us and nature. Furthermore, his values and his poet's eye offer a way back into a relationship with nature, a way back from the era of climate change, whole landscape destruction, and a sixth mass extinction – despite, or even because of, the science-led approach championed by David Attenborough and the top academics. Wordsworth understood the depth and intricacies of our connections with nature and, through his writings, almost defined nature itself. Crucially, he became one of Nature's spokespeople.

At least, that was the gist of the conversation with the Oxford undergrad I met beside the graves. We spoke only because we arrived almost synchronously to scatter rose petals there, as is the habit of some devotees of the Romantic era writers. We scatter petals on Wordsworth's grave, and on Hartley

Coleridge's grave – which lies just a little towards the church – as Hartley's father, Samuel, is buried elsewhere (at St Michael's Church, Highgate). It makes us feel better: but then, so much of what we do concerning the world of nature is done to make us feel better, to reduce the levels of metaphysical guilt within us, rather than to make nature itself feel better.

The church clock struck 7.45 during these musings: pm, and slowly, as if to slow time down, back, into the more contemplative time of the Romantic era. Midges were dancing joyously towards their deaths; the Sycamores stood aloof and broody; the sentinel Yews in pensive mood, still, composing; the air scented by lawn mowings; and a few Icelandic Poppies flowered yellow in the grey stone walls. The river which curls round two sides of the churchyard was asleep, the light slanting, the air cooling.

And at that moment, as if from out of the past, as if summer was issuing a farewell bidding prayer, a Blackbird sang, softly, whilst hidden within the Yews – the last song of summer, a sonnet.

The church at Grasmere is kept open.

Grasmere, Cumberland, 3rd August.

To me the meanest flower that blows can give
Thoughts that do often lie too deep for tears.

William Wordsworth - *Ode: Intimations of Immortality*

From Watlington Hill

You stood before me like a thought,
A dream remembered in a dream.

ST Coleridge - *Recollections of Love*

I can only visit Watlington Hill in August. At other times of
year it scarcely seems at home – absent from itself. Places can
be like that: fiercely seasonal.

Set on the Chilterns escarpment a few miles east of Oxford's
dreamy spires, the hill comes into its own when the harvest is
being gathered in on the clay vale below, and particles of
harvest dust fill the air and enhance the sense of haze, and
hazing, within a blurring summer sky. And then too, it casts
its incense, for in early August the down is carpeted in – or
rather, clothed in – flowering Wild Marjoram, which casts its
heady, soporific scent into the lower air. Wild Marjoram is the
true flower of the Chiltern Hills, not the rare and eponymous
Chiltern Gentian that grows in a few discrete spots along
Watlington's chalky slopes.

Here is where summer finds its deep fulfilment, and where
the song-dream finally settles, like a drowsy bumblebee on a
Wild Marjoram head; pollen sacks full-to-bursting. Here are
Keats's 'still more later, flowers for the bees, Until they think
warm days will never cease, for Summer has o'er-brimm'd…'
Here is England's heartland. Below, in the distance, combine

harvesters are coughing up dust, the dust of ages; but from here they are soundless, and seem placid and benign. Even Watlington's skies are surprisingly quiet, somehow: not plagued by aircraft or the roar of traffic.

There is something nationally quintessential and cathartic about the views out westwards from the upper slopes of this hill, as if all England stretches out before it. It presents a land-scape of hedgerow trees, of English Elm and Pedunculate Oak – which in August become blue-green, harmonising with the maturing summer sky. Somewhere out there, almost beyond, on undulating ground to the north-west, lies Bernwood Forest, an old Oak forest of Fallow Deer and forgotten history. It lies in a shimmering haze beneath the red beacon light of the Otmoor television transmitter, and is just discernible, for much of the time, as a vague, brooding, other-worldly darkness. A little further to the north, one can just make out the ancient hilltop village of Brill, renowned for its windmill. Away to the south-west, the chimneys of Didcot power station seem, from here on an August day, to be manufacturing nothing more challenging than fluffy white clouds: cumulus humulis, or fair-weather cumulus. The town of Watlington, itself, has scarcely changed in the 50 years I've known it. Certainly, nothing of any consequence has happened there during that time, and the place has no ambition beyond simply being itself; a place that relishes its own tranquillity.

Edward Thomas visited Watlington on his *Icknield Way* journey, published in 1913. He described Watlington as, 'a big square village of no great beauty or extraordinary antiquity, all of a piece and rustic, but urban in its compression of house against house' and relates more strongly to the pictures hanging in his inn room. He merely describes Watlington Hill as being 'thorny'. It still is.

The hill itself offers expanses of short Sheep's-fescue grass turf, studded with short downland specialists like Clustered Bell-flower, Common Rockrose and Yellow-wort, and with ant hills carpeted with Common Bird's-foot-trefoil, Wild Strawberry and Wild Thyme. People place loose flinty stones on the anthills – perhaps as touchstones, as memory traces. In August, Silver-spotted Skipper butterflies dart low over the turf, but too fast for the eye to follow. The hotter the sun, the faster the skippers speed away – and naturalists' eyes are tired by this time of year, worn out by July's stentorian days. Amongst the grey lichen patches, Striped-winged grasshoppers call shrilly, at least to people whose high-pitched hearing is still intact. In a couple of places, known only to a few, Wild Candytuft grows, but by August its flowers are long gone and the plant, an annual, appears only as dead stems offering distinctive straw-coloured seeds that tell of a drab season beyond summer.

And there are Yew woods, dark and motionless, beyond the trampled turf where children play. Only lovers enter into these extensive woods, two by two, in their moments. Perhaps the Yew woods absorb them, for none seem to return.

The lower slopes are infested by scrub, where the soil is deeper: primarily Hawthorn and Dogwood, sometimes with Whitebeam and Ash trees. Much of this developed when the Rabbit population died off, after myxomatosis was introduced. The Rabbits returned to Watlington during the early 1980s, too late to stem the advance of the Dogwood edge. For at least 40 years conservationists have been attempting to eliminate Dogwood here by cutting it during the autumn and winter months. They are still doing so, annually, but their efforts have made not one iota of difference to this enduring shrub. It likes being coppiced, which is what winter-cutting constitutes. A long while ago someone conducted a PhD into its ecology, and found that cutting it outside the main growth months of early summer is counterproductive. Nonetheless, every winter, volunteers diligently cut back Dogwood here, and elsewhere in the Chilterns. The pointlessness makes me smile, and I increasingly find myself taking the side of the Dogwood.

Everywhere, on still days, from above, high and low, comes the mewing of Red Kites, especially early in the day and again in the evening, when they gather in circling groups over this hill. Now, this is a change, and a welcome one to boot. The Red Kite was systematically persecuted to extinction in England, and sank almost from our memory. Then, beginning in 1989, it was reintroduced to suitable places in England and Scotland, including to the Chiltern Beech woods around Watlington, just as the Rabbits were returning. Ninety birds were released in the southern Chilterns. They bred there first in 1992, and at least 400 birds are now present, despite the fact that a number of young have been harvested off, to kick-start reintroduction efforts in other parts of the country.

This is one of the most gloriously positive conservation success stories, not least because the bird has contributed massively to the Chiltern's sense of identity and has become an integral part of the spirit of the place, of people's heartland, of a place where people dwell. Deeper still, it tells that much of our wildlife need not be rare, and that nature can have immense powers of recovery – when we permit it. It also tells us that we can love what we fear, or even hate: for, make no mistake, this bird was deeply, and wrongfully, vilified. Shaman-like, this bird has returned, and has greatly enhanced the place, adding a quality of ethereal durability to a landscape under immense pressure.

So, Watlington Hill still offers us a dreamscape, and by that I mean a landscape where time and place, man and nature, are harmonised; at least under the august August skies. The inner eye can wander here: northwards along the escarpment and back again, into a place which will not allow itself to be desecrated, at least during August. Here there is still the trace of some prescient dream, a state of mindfulness, which tells that only the dream itself is real.

Watlington Hill, Oxon, 10th August.

Yes, I will be thy priest, and build a fane
In some untrodden region of my mind

John Keats - *Ode to Psyche*

Dubris

We both bathed, and sat upon the Dover Cliffs,
and looked upon France with many a melancholy and tender
thought.

Dorothy Wordsworth - *Diary, 30th August 1802*

Cramped within a narrow winding valley, at the seaward
extremity of the North Downs, Dover (Roman Dubris) has
always been a place to travel through, but also a place of
intense history. It has, perhaps, become so obsessed with
history that it makes its own. Much of that history is ecological,
the pace of which must be quickening now: this is the first port
of call for species crossing the Channel by the aerial route, and
the frontline in any ecological struggle between, so-called,
native and non-native species.

Walking the cliff-top downland fringe in mid-August, your eye
is distracted by a great many things: the giant nautical goings
on mid-Channel, in the form of container ships the size of
hospitals, and cross-Channel ferries the size of multi-storey car
parks; the powered flotsam-and-jetsam of tugs and whatnots,
and the Second World War aircraft that are almost invariably
bumbling about overhead (a Spitfire today, a Hurricane tomorrow, or even a Lancaster bomber); the wealthy retirement villas
with ride-on mowers, landscape gardeners and vulgar statuary
– juxtaposed with distant views of Dover Castle on its hilltop,
South Foreland Lighthouse, the dark needle of the Dover Patrol
Monument, or the modern coastguard lookout on the top of

Foxhill Down. Natural features, such as Shakespeare Cliff (named after a passage in *King Lear*), tend to pale into insignificance. This landscape is as seriously cluttered as the contiguous sea, and the sky. Much of that clutter is history happening now…

It takes but a little while for one to realise that the ground flora – amongst the basic matrix of un-grazed Upright Brome grass and Tor-grass – is mostly of non-native origin. Garden escapees and chuck-outs proliferate here: the gardeners soon realised that these plants are highly invasive, so they eviscerated them from their gardens, and dumped them on the narrow strip of SSSI grassland. Then the gardeners planted another generation of fashionable horticultural invasives, which again got hoiked out and dumped. And so it goes on… 'A good ground cover plant', says the label in the garden centre. When will we ever learn?

Beginning in the east, at Kingsdown Leas, a 150 metre wide and mile-long length of cliff-top downland: the Broad-leaved Everlasting Pea, *Lathyrus latifolius*, dominates. Mowing during the autumn and winter months has only encouraged it to spread. There are also drifts of Goldenrod, Winter Heliotrope, the vulgar Rose of Sharon, and even a tender Mediterranean shrub called Scorpion Vetch *Corronilla valentina*, which looks like a shrubby version of our Horseshoe Vetch.

A look at the invertebrate fauna would make one half expect to find scorpions present as well. I found the purple and iri-descent-green Rosemary Beetle, *Chrysolina Americana,* numer-ous in the grassland. This is a recent arrival to the UK, a pest of aromatic garden herbs. It may be feeding on indigenous

Wild Thyme on Kingsdown Leas, rather than on any of the invasive garden plants that are taking over the cliff-top grass-lands. It may even have colonised naturally, from across the Channel – rather than as an accidental hitchhiker in an im-ported container-grown plant, the normal route. The first male Ivy Bees were starting to appear, in synchrony with their beloved Ivy flowers. This distinctive stripy bee was first recorded in the UK in 2001, on the Dorset coast, having spread naturally from the continent. It must have colonised the Dover cliffs around that time, if not earlier. I was also bemused by a paper wasp, with long dangling yellow legs. This should have been *Polistes dominula*, itself a recent colonist to the South East from abroad, but other, similar-looking species occur close by in France – I should have taken a net with me and retained the specimen, as it might have been a species new to Britain. Opportunity lost.

Kingsdown Leas slumbered dreamily above the Dover Strait until the halcyon summer of 2013, when it was hit by the vanguard of an invasion of the Long-tailed Blue butterfly on a scale unprecedented in entomological history. Previously, this global wanderer had occurred only as a rare vagrant to the shores of Albion, but that summer it conquered much of coastal Kent and Sussex. Its larvae feed up in the pods of a diverse range of plants of the Pea family, seemingly favouring Broad-leaved Everlasting Pea. Kingsdown Leas proved to be a paradise for it. The butterfly returned in 2015, though in lesser numbers, and was rumoured to have put in an appearance in 2016. But on this trip I could find no sign of the distinctive butterflies, which flit and dart like a Hairstreak rather than bumble around like a Blue, the tiny white eggs on the sepals, or even the hatched egg cases on developing pods. I was too early. Two days after my visit, though, the wind changed to the south-east and might have escorted the butterfly across the Channel. Success in natural history depends greatly on being in the right place at precisely the right time.

The Long-tailed Blue was not the only Lepidopteron invader to turn up here recently. A friendly lady told me that her cat, a black-and-white minx called Poppy, had recently brought a giant Death's Head Hawkmoth in through the cat flap, and released it into the kitchen to generate maximum chaos by leaping wantonly after it, whilst dinner was being prepared. Hawkmoth and cat were eventually both removed, relatively unharmed. Later, I met Poppy swishing her tail on the garden wall. She ignored me, even after I'd congratulated her on catching a rare migrant moth that I have desired to see since childhood but have yet to encounter.

Between Kingsdown and the Dover Patrol Monument, which stands on the summit of Bockell Hill, the fringe of cliff-top grassland narrows, forced seaward by harvest fields that seem to stretch as far as the Channel itself. Attempts to widen the fringe, by taking a strip out of cereal production and setting back the fence, have resulted in a riot of plants wrongly accused of being 'arable weeds'. They are not weeds, but the gay cornfield plants of yesteryear, forced into rarity by modern herbicides. Impressive here was an array of indigo-blue Basil Thyme, a scarce plant of disturbed chalk – more of it than I'd seen in a lifetime of downland wanderings.

Looking back along a line of 40-metre-tall chalk cliffs, from a minor promontory, a few Fulmar and Kittiwake were flying close to the cliff, keeping below the radar. Their populations here are weak and diffuse, but linger on. Ravens have recently reappeared along the White Cliffs, after an absence of many decades; whether from further along our coastline or from France is anyone's guess.

Past Bockell, much of the vegetation growing on the narrow strip of cliffs at St Margaret's at Cliffe again reflects horticultural fashions impacting on the native chalk downland flora and fauna. Besides an abundance of invading Buddleia, Rose of Sharon, Shasta Daisy, Goldenrod and Lady's Mantle *Alchemilla mollis* are various plants that are not normally invasive, but have become invasive here in the Kent coast's hospitable climate: notably the Narrow-leaved Everlasting Pea, *Lathyrus sylvestris,* which drifts about over the down in huge clumps, and which flowers, pathetically, for all of a fortnight. Worse, most of these non-native nasties are unpalatable to the docile black-and-brown Dexter cattle that are pastured here for conservation grazing. This means that the cattle readily graze down the palatable native herbs and grasses, and so enable the untouched non-native garden thugs to spread further, by removing the competition. All this was mumbo-jumbo to a local lady, a lady of no small consequence, who had demanded to know what I was doing. "We like it as it is!" she expostulated, then huffed and puffed away, grumbling at her tardy dog, an obese three-legged Jack Russell: "Come on, Bomber Harris!" she shouted at it. Noel Coward, who lived here until the village filled up with hoi polloi, would have relished her cameo appearance.

And that's precisely the problem. People do like, and indeed love, our countryside as it is – believing erroneously that it has always looked like this, and ever always will. They do not notice the constant, insidious, erosion of natural features inflicted by each technological advancement, agricultural development, fashion, or policy change; let alone the subtleties of natural change. Worse, few have the imagination or the descriptive vocabulary to perceive what our countryside used to look like, during more benign eras, let alone how wondrous it could realistically be with just a little less orc-like behaviour on our part – we were taught simultaneous equations, and GCSE French, instead. So each generation settles for more and more homogenisation and mindless buggerisation, and less and less wildlife – and in consequence our relationship with nature continues to spiral downwards. Edward Thomas's rural prose describes an abundance of wildlife beyond our comprehension, including Wrynecks, from just over a century ago.

By the time I'd reached the lower, eastern, end of Lighthouse Down I was cross; and on a day of mare's tail cloudlets wisping in the bluest of skies, with the cliffs of La Belle France dancing in and out of focus in the distance. The dog woman had started it. Lighthouse Down is backed by only one or two villas, and consequently supports a far richer, and more native flora, with fewer garden escapees or chuck-outs. Sure enough, though, two wretched species of Cotoneaster are invading the precious thin soil, short turf, areas with malevolent intent.

The Chalkhill Blue butterfly restored my sense of perspective. Clothed in Cambridge blue with a hint of iridescence, the males were busily searching the short-grass breeding grounds for the freshly emerged, sepia-brown females; and instantly having their way with them: no questions asked, no introductions needed, no name-asking – just instant mating, for upwards of an hour. Mating pairs were strewn about all over the place, wings closed: rather like undergraduates after a May Ball. They'd earned it, having endured the traumas of caterpillarhood and pupation.

And there, in her web, amongst a patch of purpling Greater Knapweed heads, was a female giant Wasp Spider, *Argiope bruennichi*. This impressive wasp-striped beast was first recorded in the UK in 1922, at Rye in East Sussex, and so may have been resident here along the Dover coast for almost a hundred years. It feeds mainly on grasshopper nymphs, and likes its grassland rough, uncut, and un-grazed.

South Foreland Lighthouse, as it stands, was opened in 1842, and is whiter than the white cliffs it stands upon. In sunshine, it dazzles blindingly. Now technically defunct, it hosts one of the best tearooms in England: one capable of rivalling The Orchard at Grantchester, outside Cambridge (established in 1897). Mrs Knott's Tearoom is named after George Knott, a lighthouse keeper who lived here with his wife and 13 children. The tearoom is a recent invention by an inspired National Trust manager, offering loose-leaf tea in a gloriously random assortment of china tea pots, accompanying an eccentric mishmash of bone-china cups and saucers. If you have a bizarrely-coloured china teapot, or cups and saucers you can't live with and want to find a good home for, bring them all here. The cakes and scones are, of course, par excellence, and central to English culture. Tourists walk two miles along the Dover cliff tops to visit this best of all possible tearooms.

Refreshed, for cake is restorative, I was heading in the opposite direction, west, past drifts of Wild Carrot growing where narrow cliff-top fields had recently been taken out of cereal production and allowed to revert naturally. These ex-fields were going through their Wild Carrot phase, but will soon enter another phase as the process of succession and colonisation continues.

Fan Bay is an oddity, a corrie-like combe in the chalk, directly behind the cliffs. Its slopes offered another Chalkhill Blue colony, though mating was over for the day and the males were instead basking in late afternoon sunshine, knackered most probably. The combe possesses a brace of recently restored concrete sound mirrors: pre-radar technology for determining the approach of enemy aircraft, first set up in 1917; and a Second World War deep shelter, where battery gunners could hide during bombardment. This late afternoon, Fan Bay was vibrating with the songs of the Long-winged Conehead, a southern cricket of long vegetation whose stridulation sounds like the crackling of a pylon line in drizzle. The entire combe was merrily crackling away. One wonders what the khaki-clad technician listening out with a stethoscope, by the sound mirror, would have made of it, but then the conehead probably wasn't here back in 1917.

Westwards, and onwards towards Dover, lay the pony-grazed downland combes of Langdon Hole, and the huge old gun emplacement terraces on Langdon Cliffs above Dover harbour, now a vast tiered National Trust car park. The place was swarming with people of all nations, come to ogle at the port activity, listen to the tannoy announcements in three languages, and ingest the fumes that readily surge up the cliff face. It was also bouncing with Cabbage White butterflies, both Large and Small varieties; but then Dover is the Cabbage

White capital of England. Either the butterflies swarm over the Channel to here, or they breed profusely on the tender young leaves of the abundant Wild Cabbage plants that grow on the bare cliffs, or both. They were gathering to roost on Wayfaring Tree foliage, on which they are superbly camouflaged.

Dover is a place of arrival and departure. Julius Caesar tried to land here. Lord Byron spent his final night in England at Dover's Ship Inn, fleeing from catatonic scandal which reverberates today. Everything and everyone ends up at some time or other in Dover: it's that sort of place.

St Margaret's at Cliffe, 20th August.

Listen! You hear the grating roar
Of pebbles which the waves draw back, and fling,
At their return, up the high strand,
Begin, and cease, and then again begin,
With tremulous cadence slow, and bring
The eternal note of sadness in.

Matthew Arnold - *On Dover Beach*

Echoes of the Long Hot Summer

So strong was the young beauty of the year, it might have seemed at
its height were it not that each day it grew stronger. The new day
excelled the one that was past, only to be outshone by the next.
Day after day the sun poured out a great light and heat and joy over
the earth and the delicately clouded sky. The south wind flowed in
a river straight from the sun itself…

Edward Thomas - *Light and Twilight*

Pine cones stood out proud, flayed rigid by weeks of dry air.
Below the burdened trees the needles lay desiccating, on
parched, bleached, ground where the A272 departs from the
hectic A3, just to the north-east of Petersfield. On the right,
southwards, the great downs of Harting, Round Down and
Beacon Hill appeared grey, but drifted in and out of focus in
shimmering heat, under a sky that would have been cerulean
blue had it not been for the shroud of heat, hanging dust and
myriad microscopic insects. Close by, the purple heads of
Tufted Vetch were the only flowers to add colour to the dead
straw-like stems of False Oat-grass, which had grown stunted,
and then wilted, weeks ago. Drought was omnipresent. The
hedgerow Elders had aborted their flowers, and would set no
fruit this summer. Now, their leaves were curling and turning
colour, soon they would drop.

Further along, on the south-facing greensand slopes towards
Rogate, acres of maize and potatoes lay wasting. Huge swathes

of these crops had failed, revealing burnt-umber earth that reflected heat back into an already over-heated sky. Durleigh Marsh had long run dry, and the cattle, short of grass, had grazed it right down, rushes and all.

In Rogate, Nasturtiums in window boxes had shrivelled up, the church bells had fallen silent, and the White Horse pub was advertising stripper nights. Cars snailed past – Hillman Imps and bright new Morris Marinas, reflecting sunlight, caught behind an ageing Morris Minor which crawled sensitively through the heat. The tarmacadam was melting, again, such that the tyres of passing vehicles – and even bicycles – offered a curious suction sound.

Outside St Peter's Church, in the hamlet of Terwick, the Lime trees had dropped their paltry seed crop; and their leaves were covered in dust which had adhered copiously to thick layers of honeydew – the sticky secretion of that summer's aphid swarms. And here, in the hollow, dust clouds hung, seeking something to settle upon. Deep in the shade, a family was picnicking, the doors of their shiny new Ford Cortina open, radio blasting out *Fernando* into the 30 degree ether.

'There was something in the air…' the voices sang. Was something wrong with the sun? If anything, it had become too strong; and the very movement of air had been stilled by the heat.

At the foot of Cumbers Hill, the journey's first stand of dying English Elms offered premature tints of mid-autumn, in early August. Further on, towards Trotton, more roadside Elms were dying, for this was the summer wherein Dutch Elm Disease flared up rampantly. The Rother, at Trotton Bridge was sunken, stagnant, a dull green pool – covered with a lifeless film of dust on the west side, and to the east a pebble bed through which tepid dirty water seeped. It had become a series of unconnected pools, no longer a flowing river. The Dippers had gone. The Grey Wagtails remained, but fretted.

On Iping Common, on the Lower Greensand ridge, the Birch leaves had yellowed, and were falling. Some Rusty Sallows were dying, and the Bracken had turned colour, and flopped. One large area of open heath lay blackened: the site of one of the summer's numerous heath fires, now quenched. The Birch drought continued past Severals Common, and all the way to baking Midhurst. On the verges, only Chicory and Mugwort offered token flowers to the traveller.

In the town, the scarlet Geraniums outside the crossroad convent, opposite Russell's Garage, had been kept well-watered by diligent nuns, though the content of the hanging baskets in the High Street had frazzled into something long-dead, and indeterminate. Vestiges of Ragwort and Purple-loosestrife flowers lingered on the marsh below the ruins of Midhurst Castle, but water had ceased to flow over the Rother's weir spillway, and in the neighbouring village of Easebourne the Hydrangeas had dropped their foliage, leaving colourless flower-heads – like something out of a dried floral arrangement.

Beyond Midhurst the drought intensified, within the awesome beauty of that noetic summer: a group of great veteran Beech trees stood dying, their limbs reaching, out, to a remorseless sun, the Bracken lay bronzed. Benbow Pond was a morass of cracked dry mud; the fish gone, rescued. The normally verdant fairways of Cowdray Park Golf Club stood grey, matching in colour the Bent grass roughs. This dry acid grassland was alive with Field Grasshoppers and Small Copper butterflies.

Then, past Halfway Bridge – a bridge over a sludgy trickle – the barley had been harvested; miniscule grains that had failed to swell. The wheat, little better, was now being cut, which meant that more dust was being cast into already-laden air. Soon straw burning would add to the heat, and to the brown pall around the sun. Burning had started already, away on the South Downs beyond Bignor Hill and Amberley Mount, for plumes of grey-brown smoke were rising upwards, before bending suddenly on a high-level south-easterly, and drifting out across the Low Weald. Two fields further along the road, the combine harvester had broken down, apparently due to the plethora of aphids and ladybirds – a familiar problem to the South's farmers that summer. The road flowed past it all, leaving it in a cul-de-sac of time, to become memory.

Past the clustered village of Tillington and another stand of dying roadside Elms, their upstretched limbs curling in at the tips in abject surrender, plants growing in the high stone walls of Petworth deer park had long died off, and the grey parkland grasses had been grazed right down. In the town itself the roads had become viscous, where the sun had been funnelled down and heat had reflected downwards off its tall houses. Tar stuck to people's shoes. The Sparrows were quiet, bathing in dust beneath a great Red Cedar tree. A bevy of Small Tortoise-shell butterflies sought out the panicles of a Buddleia bush overhanging the great wall of Petworth House.

On the other side of Foxhills' wooded Greensand Ridge, the

road finally ran down on to the Weald Clay. Crumpled Hazel leaves littered the verge through the dense Oak woodland of The Mens, where the Honeysuckle leaves had also wilted. The dusted windows of a lonely red phone box on the treed corner at the hamlet of Strood Green were cobwebbed over, and the concrete floor littered with dead ladybirds. The open fields here were cracked and deeply fissured, for the parched clay had shrunk into hexagonal plates, which wobbled if one stepped on them.

Wisborough Green's village green was now firmly brown, its cricket pitch cracked – a spin bowler's paradise. Stoolball was being played on the outfield, almost in slow motion, as the heat had slowed the nation down. Opposite, the old Zoar Chapel (founded in 1753) had seen nothing like it, at least not since the great but forgotten summer of 1893, when the sun had burnt remorselessly from March to September. The brick-lined village duck pond had dried out, but its ducks remained happy, for ducks love all weathers. Then, before the village's end, the row of mighty English Elms, which for centuries had shrouded the church of St Peter Ad Vincula (St Peter in Chains) up on its grassy hillock, was shot with dying branches. The giants stood there, along the foot of the mound, proffering a motley mix of deep green leaves and yellowing, browning branches. At their death the village would change for ever. Swallows skimmed idly by, oblivious, in the thick air.

Before Billingshurst, a broad grassy bank sported Common Knapweed, Common Ragwort and Common Restharrow flowers, and a thriving colony of Wall Browns – a heat-loving butterfly which thrives in hot summers. Billingshurst itself was announced by stands of yellowing Elms, and ended with a garden full of apologetic Hollyhocks, offering a few flowers on half-grown stems.

The bank leading into Coneyhurst was a vast yellow swathe of Tall Melilot, which somehow intensified the aura of golden light – for this summer sought goldenness, from out of azure. It hummed with bees, and pulsated with grasshopper song. A giant Elm near Coolham crossroads stood proud, deeply green, resisting fungal infection. A pair of ragged Carrion Crows, in full moult, hung high in the shadows of its branches.

Turning left for Dragon's Green, at Shipley crossroads, you might pass a young man, heading for the woods, having hitch-hiked along the A272, and written down what he had witnessed. His long hair was bleached by the sun, his nose full-reddened and peeling, his feet hardened by weeks of walking on ever-hardening ground in cheap Spanish fell boots, and his flared jeans had faded in the intensity of sunlight. His knapsack contained a towel, a spare shirt and socks, binoculars, a butterfly net, some inappropriate smoking material, a volume of unreadable poetry, a notebook and several pens. He was walking through paradise, and had been doing so for months – all summer. Above him, a pair of white butterflies met, randomly; they spiralled upwards together, in a courtship dance, and ever onwards into golden light. He would never forget them: the least of all the myriad butterflies he witnessed during that most magical of all summers. This was the long hot summer of 1976, and he would spend the rest of his life striving to return there, trying to find a way back into a life that time had forced him to leave behind. Distant, from a tangled sloe hedge, a Turtle Dove purred.

Dragon's Green, West Sussex, August.

I was brought into this world,
I did not come here alone.
Whoever brought me here,
Will have to take me home.

Jalaluddin Rumi - *Whoever Brought Me Here* (translated, arranged)

Ladybird Days

Ladybird, ladybird fly away home

Anon - *Ladybird, Ladybird*

Perhaps the most remarkable natural event of the long hot summer of 1976, and one of the best remembered, was the great ladybird invasion. One observer, in Louth, Lincolnshire, recalls: 'I remember seeing the greenfly and aphids drifting past the doors like green smoke. This went on for at least a couple of days. Then came the ladybirds!' They were tracking the aphid swarms. Ladybird populations tend to respond strongly to prolonged fine summer weather, reacting rapidly to increases in the aphid populations on which both the adults and larvae feed.

The 1976 influx, which climaxed for a month from mid-July, probably engaged the great British public more than any other wildlife event of the 20th-century – for there was no avoiding the ladybird swarms. There are accounts of holiday-makers running from beaches in Brighton and Weston-super-Mare, and some spectacular evacuations occurred at the Butlins holiday camp beaches at Minehead, Pwllheli and Skegness. The ladybirds were biting people, apparently: or maybe people, or the ladybirds, had been watching too many Hitchcock movies. All the starving ladybirds were seeking, of course, was moisture and food. They were desperate.

Testimony: 'On 11th July 1976 I was piloting a light aircraft about 20 miles south of Manchester at 1500 foot when I flew into a large swarm of ladybirds. It was like flying into bird shot.'

Testimony: 'My daughter and her husband reported sailing through seas that were completely covered in ladybirds!'

Testimony: 'It was just before lunch, and the whole sky went black…'

One newspaper headline read, 'Ladybirds Take Brixton!' Best remembered is footage from the BBC1 weekday live news magazine programme *Nationwide* of two somewhat Monty Pythonesque ladies fending off wild and dangerous ladybirds with their handbags. The ladybirds won.

The ladybirds appeared to be attracted to the colour yellow. A yellow Hillman Avenger became covered in them whilst parked on Cleethorpes Promenade, and turned scarlet, though the account states that 'most had gone by Doncaster' on the way home. On Hunstanton Promenade, a broad yellow line was painted deliberately to attract ladybirds, which could then easily be removed by road sweeper. That line remained for years, a testimony of a summer of true greatness, a symbol of our struggle against the power of nature.

Some science is needed to explain this mayhem. It is provided by the late Professor Michael Majerus, Professor (of ladybirds) at Cambridge, in an article in the Journal of the British Entomological & Natural History Society (known in entomological circles as the 'Brit Ent Soc'), published in 1996 and co-authored by his wife, Tamsin. The 1976 influx, they write:

'was truly astounding. It extended across more or less the whole of England and Wales, and into some parts of southern Scotland. Exceptional numbers of ladybirds were also recorded in many parts of north-west Europe.'

The reasons are complex. Ladybird population explosions, the account explains, are stimulated by food availability, a mild or steady winter, sunshine in early summer, high midsummer temperatures, and, importantly, the absence of ladybird predators and parasites. These interactions are in turn dependent on climatic factors, for populations of winged insects are heavily influenced by weather – booming during periods of fine weather, and busting during poor summers. The roots of the 1976 swarms lie back in the hot summer of 1975 when, as the authors explain, populations of Two-spot, Seven-spot and Ten-spot Ladybirds boomed and produced rare second broods. The beetles then survived well during a mild winter (the winter mortality rate is normally around 50%). Crucially, this was probably happening all over north-west Europe.

Then, aphids abounded early in the summer of 1976, whilst populations of the two Phorid flies (*Phahcrotophora fasciata* and *P. herounemis*) which parasitize ladybird pupae, and the Mirid bug *Deraeocoris ruber*, a major predator of ladybird pupae in Nettle beds, were undoubtedly low. This meant that the host (the ladybird) had leapfrogged ahead of its parasites and predators. All this was a perfect recipe for a ladybird population explosion. But it could not last. The account continues: 'Bluntly, the aphids were massacred ... [and] the plants the aphids were feeding upon were deteriorating rapidly under the scorching heat.'

By mid-July vast swarms of ladybirds – one scientific estimate puts it at 23.65 billion of them – had taken to the air, and were being carried around by light winds, which chopped and changed direction almost daily, being effectively random in strong anticyclonic conditions. Many swarms were brought to ground by air currents along the coast: they were dumped by thermals whilst travelling perhaps a thousand metres up; thus, many of the coastal dumpings were probably of home-grown ladybirds, rather than of migrants from distant countries, though it is likely that major influxes also occurred.

Recently, another theory behind the great influx of 1976 has been produced, by Simon Leather, Professor of Entomology at Harper Adams University. In 1972 a new wheat variety, called Maris Huntsman, was introduced. This high-yield and fungal disease-resistant variety accounted for a third of the winter wheat grown in the UK in 1975 (though it was subsequently found to produce poor quality bread). It produced good thatching-straw, and aphids loved it. Perhaps this winter wheat variety was a contributory factor behind the great ladybird swarms of 1976, for a single ladybird can consume 60 aphids a day, and the Maris Huntsman wheat fields were full of aphids.

The 1976 population explosion, which consisted mainly of the common Seven-spot Ladybird, seems to be the greatest on record in the UK – but by no means the only one. There is an account of a major influx in east Kent in *The Times* of 15th August 1869: 'In Ramsgate men were shovelling them down the large grating into the sewer... the streets seemed covered in red sand. The people coming out of St George's Church, Ramsgate, at 10 minutes before 1 o'clock on Sunday were covered with the insects before they had proceeded many yards.' Similar events had been observed there in mid-August, 1847.

The last great ladybird year prior to 1976 was in 1959, and the next subsequently was the hot summer of 1989 – only this was an order of magnitude less, and populations subsequently collapsed during the summer of 1990 when the parasites and predators caught up with the host.

In recent years, populations of the Seven-spot Ladybird have been in the doldrums, perhaps due to the impact of modern agricultural and horticultural pesticides on aphid populations all over western Europe. There have been only a few local out-breaks, most notably in 2009 – but none anywhere remotely near the standard of the 1976 epidemic.

Nowadays, the ladybird most people see in the UK is the Harlequin Ladybird, *Harmonia axyridis*, a new colonist which was first recorded here in 2004. This aggressive, non-native, invasive species has now colonised nearly all of England and is making inroads into Wales. It may be out-competing or pre-dating some of our native ladybirds. Back in '76, however, the Harlequin Ladybird occurred no nearer than Kazakhstan. Much has changed in our environment since then. Wildlife can be hugely dynamic, especially insect populations which can undergo population booms on a scale almost beyond our comprehension.

But without the Harlequin, which is primarily a tree-feeder and which produces its main emergence in early autumn, most of us would scarcely encounter ladybirds nowadays: our so-called common native species are that badly depleted. It is quite possible that the next major outbreak of ladybirds in the UK will not be of a native species and it may happen in early autumn, should the Harlequin produce a massive brood after a long hot summer – should the long hot summer ever return.

Above all, this remarkable tale illustrates how incredible the long hot summer of 1976 was, and suggests that many of our warmth-loving insects could be far more profuse than they are today, given suitable conditions.

Culkerton, Gloucestershire, mid-August.

Heat stretches the day

Jonathan Davidson - *Summer, 1976*

The Triumph of the Skies

> But how the swifts rage!
> With slash of scored sky
> they write summer's page.
>
> Alison Brackenbury - *Postcard*

Even though the Swifts had departed, the August skies were full of reeling birds – at least intermittently; for at times the entourage of House Martins and Swallows, young and old, would speed off elsewhere, venturing far, pushing boundaries, but always returning towards eventide. They understand root-edness and belonging, even though there are few, if any, limits to their aerial ambitions.

The Martin season had begun in memorable fashion, and remarkably early too, in our sleepy and rather nebulous village on the Cotswold edge. Way back in time, on 4th April, I had been woken at dawn by ecstatic babblings coming from out-side, above my east-facing bedroom window. Two weeks too early, I thought, and rolled over to listen to Percy Bysshe, our upper-garden cock Blackbird, conducting matins, high in the leafing Bramley tree. The chortling came again: once more I denied it – I was hallucinating spring again... but then one should hallucinate spring, if only because it helps us through winter; but it is wrong to attempt to rush spring, so a meditative slumber beckoned once more.

Then the soft-and-subtle sound came a third time: my cat, Lottie The Berserk, reacted as only a lunatic tabby-and-white could – by forsaking the bottom of my bed and climbing up the window pane, chattering loudly, and swishing her tail. She is a useless huntress, but entertains a ridiculous fantasy about catching a House Martin – mission impossible for even the most adept of feline hunters and a proscribed activity in my domain. A pair of Martins had indeed just returned, after a pilgrimage of perhaps 6,000 miles, and were understandably overjoyed to find that the nest they had constructed (or perhaps been born in) during the previous summer had survived a wet, windy, and thoroughly malevolent winter. They were making poetic speeches about it – either that, or they were negotiating with the group of Wrens who had been winter-roosting there, or both: it was hard to tell, especially within the cacophony of an April dawn chorus.

The two pioneer birds then had to brave the vagaries of variable early April weather, desiring to roost at night in the nest, where they often burbled in their sleep; but disappearing – presumably southwards – during a lengthy cold spell. In mid-April another pair arrived, again at dawn, and again to find their nest intact – only this time Lottie got over-excited and fell off the window sill, then ran off with a fluffed-out tail.

More Martins arrived, in pulses, over the next fortnight; some to old nests, others, presumably younger birds, to splat long lines of mud beads under the eaves, until a base for a new nest was established. There was a plentiful supply of mud this year, though the birds only gathered it on certain days. Maybe they use some avian equivalent of Rudolf Steiner's biodynamic calendar for agriculture and horticulture; deeming that today is a mud-gathering day but tomorrow isn't, so you must then wait until Sunday, which is the next designated mud-splatting day. New nests were made, and damaged nests repaired, using two main mud colours – a mid-brown and a chalky-grey-marl. Some new nests were wondrously bicoloured, and changed hue with the weather: looking umber and dove-grey in the wet, but mummy-brown and bright-buff in dry weather. Each nest takes about a thousand mud beads.

By early June the activity was frenetic. The House Martins entranced most of the 200 or so visitors who had viewed our garden when it opened under the National Garden Scheme, one azure Sunday. The Martins, of course, ignored the human visitors; but then, they ignore humans altogether. They have other priorities – small flying insects, mud, vertical walls, feathers, gaping mouths to feed, and probably both internal and external parasites, not to mention the weather.

In mid-June the first young became vocal: a squeaking that steadily increased to a near-constant chattering, under loud instruction from their parents. It persisted all night. House Martins are noisy and decidedly mucky neighbours, but then there are two sides of every coin (and an edge), especially in nature. Our problem is that we favour single-sided coins, and don't do edges at all (call the edge 'spirituality' – the thing that links the two polarised sides). Piles of poo last long into the winter on our windowsills, and attract a species of small brown beetle which one day I may manage to identify.

In early July some of the nests exploded, as over-contained teenagers finally burst out. These young returned to roost in the broken nests at night and for nocturnal burblings, bedtime stories perhaps. Their parents chose not to repair the nests and start a second brood, presumably they were discouraged by indifferent summer weather. Meanwhile, young in the new nests were becoming vocal. All told, our house had been encrusted by 25 House Martin nests: a 25-year record and a veritable blessing.

Some pairs raised second broods, but during August most nests were occupied only at night, and quietly. The Martins steadily became birds of the air, ruling the village skies, gathering for evening confabulations with their Swallow cousins; and chasing innocent vole-hunting Kestrels and marauding dark-hooded Hobbies out of the parish. Theirs was the triumph of the late summer skies.

But Oh, that vast ethereal August cornflower sky! – populated by myriad airborne insects, invisible to us, drifting about on levels unseen, and by cirrus clouds composed of ice crystals and forever hazing off into the far distances. Oh, to disappear, to lose oneself in that vast magnificence! Oh, to be able to revel in it! – like the Swallows and Martins.

Soon the Martins would be gone. Already I was starting to miss them, even as their numbers and activities were at peak. Their departure would constitute an emptying of the sky. Our love of nature is not without pain, much of which is incurred by seasonal change.

Cotswolds, Gloucestershire, August Bank Holiday Monday.

you have no time, hovering
attendance on a gallery of throats
gaping for more; their eavesdroppings
writhe whitely down our overhang,
fetching subsistence from the sun
or slipstreaming through the osiers
with kissing sip, you do what you must
all summer long – for Joy, it seems.

Keith Chandler - Swallows

The Final Ascent

Gone, gone again,
May, June, July,
And August gone,
Again gone by.

Not memorable
Save that I saw them go,
As past the empty quays
The rivers flow.

Edward Thomas - *Gone, Gone Again*

Leaving the ancient coach road behind, a place of painful noise, to follow the old sunken track up, along the edge of wheat half-cut, as the day's first Cabbage White took to the air, it seemed that summer wanted to linger on the hilltops, especially here, on Cherhill Downs, where the year's song-dream had been born. Go higher up and higher in, the day called. Early morning clouds were once again dissipating here.

The Hawthorn blossom, which had abounded here, had been replaced by a mass of reddening berries. The Fieldfare hordes would not starve here, at least for a while. But below the thorns, in the grass-flushed bank, as pristine as freshly fallen snow at sunrise, lay myriad pink-flushed blooms of Lesser Bindweed, each opening in soft morning air, to be tended late into the August day by small furrow bees (of the *Lasioglossum* family) and the common curious-looking long-snouted orange-and-black hoverfly that hovers under the unedifying name of

Rhingia campestris (it deserves a richer name).

The path had in-filled, narrowed down by rampant growth of Cock's-foot and False Oat grasses, studded here-and-there with Common Agrimony seedheads, and everywhere nettling over, after a warm wet summer in which vegetation growth had over-exceeded itself. The wheat field edge was all but empty of arable flowers, consisting of a narrow herbicide spray-zone called, unappealingly, the sterile strip – a term that sums up much about modern farming: perhaps there is even a farm called Sterile Strip Farm somewhere, there should be many.

In a shady section two Speckled Wood butterflies danced a circle dance, in a spider-webbed sunlight shaft, beneath Crab Apple trees bent over with fruit and dense Ivy entanglements. A Harvestman Spider dropped down from above to crawl wantonly across my arm. A veteran Beech hung heavy with nuts, so profuse it seemed the tree itself had changed colour from leaf-green to nut-brown. It would be a mast autumn: the Pigeons would grow fat. The Blackberries, though, were late here: red and green, hard – for summer had been poor.

Downslope, along the chalk path, discarded leaves and fallen seedheads had been washed into tiny heaps by a recent rain-storm, deposited where water had eddied. The path surface ran hard and slippery, polished by pulses of summer rain. It was not a morning for hurrying.

Now, here, past Guelder Rose heads, hung heavy with blood-red berries, succulent yet blatantly distasteful to us, suggesting plastic, the path opened into a deep gully of chalk grassland bedecked with tall herbs – Field Scabious, Common Knapweed, and drawn-up greyed heads of Kidney Vetch, long gone to seed. The Common Furrow Bee, *Lasioglossum calceatum*, sombre red-and-black males with long antennae and distended narrow abdomens, waited for the sun's return from behind a passing cloud.

No Lark ascended in ecstatic song: spring's music had long gone. Instead, from deep amongst the waving matted grass, on both sides, as the sun's warmth drew the day's first beads of forehead sweat, grasshoppers called loudly and incessantly – the music of summer's finality. This was the hour, and the day, long. It seemed the ground pulsated, and almost heaved, with their rhythms – Field and Meadow grasshoppers, profuse, all males, all calling, in need, and with urgency. And on the hill-top a wooden gate, slightly shrunken by summer sun, rattled its own song-dream to welcome one who loved this place.

Southwards lay Calstone Coombes. They had deepened and steepened, it seemed, in summer heat. Even here, on these thin chalk-soil slopes the grasses had grown unusually matted and dense in a wet summer, though now they were grey-bladed and browned-stemmed, ageing. Where the turf was shorter it was studded white with Eyebright, blue with Harebell and Devil's Bit Scabious, and wine-red with Saw-wort and the last of the Common Knapweed. June's Fragrant Orchid stems stood out brown and stark, winterine.

A meagre brood of Adonis Blue butterflies, all that the mean summer had allowed, flashed through standing stems, from flower to ageing flower; the monuments of summer passing. A lone Clouded Yellow had arrived, to patrol, at pace, the bottom of the south-facing slope of the main combe, with wings of living gold, a lone and frenetic male in a valley void of females. The grass had grown too tall for them this year, and the herd of Aberdeen Angus cattle had not been able to keep pace with its growth. Instead, there were robberflies amongst the grass tussocks, mean and grey, and the ground was littered with Heath Snails, *Helicella itala* – more a downland species than a heathland one, so rather misnamed. In places these flattened, whorled, black-and-grey banded snails scrunched underfoot. In hot, dry summers they bury themselves in the soil; but in damp summers, such as this, they lie about on the surface, bathing. It felt as if I was trampling on summer itself.

Sudden, even as a flurry of Meadow Browns spontaneously ascended, came a slow but distinct ticking, like the sound of mother's old Singer sewing machine, getting going to make something in the run-up to Christmas. Tick   Tick   Tick Tick   Tick   Tick   Tick Tick Tick Tick, and faster, till the gaps between the ticks all disappeared: TickTickTickTickTick and on, in a constant stream. It came from a tussock somewhere up-slope, but the vocalist was almost impossible to locate – as the song, or stridulation to be precise, is too non-directional to trace, and deeply hypnotic. The caller goes under the glorious name of *Decticus verrucivorus* – pure onomatopoeia, with a hint of assonance – the Wart-biter Bush Cricket: a giant of an insect, a top national rarity restricted now to a few southern downs. The song's meaning was simple: time is running out here, it is slipping away, now are the final oozings of summer hours; and finally, remember me.

Cherhill Downs and Calstone Coombes, Calne, Wiltshire, 31ˢᵗ August.

Crickets ticking in the long grass know
what we don't. There's not much time.

On this day without a point we lie
and watch the clouds go by.

We can't live like clocks and feel
the ache of every stroke.

Katharine Towers - *August*

Wingletang

Here love ends,
Despair, ambition ends,
All pleasure and all trouble,
Although most sweet or bitter,
Here ends in sleep that is sweeter
Than tasks most noble.

Edward Thomas - *Lights Out*

Like the grave, the sea calls us all: even those who are children of the mountains or meadows, or, like me, of the forest – whose lives are acted out almost beyond the call of gulls or the tang of salt; almost. Yet that amorphous, lachrymose, mass of chemistry and biology, which seems lifeless and yet is life-full, life-giving and life-taking, holds some mighty sway over all of us. It responds to the pull of the moon, the rise and fall of atmospheric pressure, the seasons, sunlight, the colours of the sky, serenity, and the hurricane winds, yet it is its own master.

On Tresco, where the constant squabbling of House Sparrows renders a sub-tropical garden wholly English, low morning cloud and the faintest drizzle on the trailing edge of a passing warm front, lifted suddenly into an afternoon that seeped September. The island became golden-ised, scarcely of this world. Small Copper butterflies basked on yellowed sands, as miniature fire dragons. The sea became aquamarine. It was one of those afternoons that seemed born solely to create the perfect sunset, at the lowering of August's sun into September's equinoctial skies. All around the Isles of Scilly people were settling down to watch: for here people dwell close to nature, and watch, learn, and know.

Yet, with an hour or so to go, thin cloud spilled over from the Western Approaches, steadily turning the developing sunset from blood-red to silver, before becoming lost in an ominous leaden-grey. Perhaps it was the probing, shoreline Turnstones that did it, for they became more frantic as the sun lowered itself. Perhaps they deliberately or inadvertently turned some secret key? Certainly, it was as if the evening had been flipped over, from rosy to grey, from sun to rain – as in the chance turning-over of a small pebble, a keystone.

The greyness was the advancing edge of an ex-tropical storm, downgraded and downsized, but the first of the autumn gales, a precursor of things to come. Its rising wind forced the Swallows to hunt low along St Mary's narrow, sunken tarmacadam lanes, where small insects flew naively in the lea of the stonewalled Cornish hedges. A wetting drizzle set in – fine but saturating, onward-driven by the wind.

It forced me to seek shelter in a bird hide at Higher Moors, behind Porth Hellick cove, on the southern flank of St Mary's. Bird hides are usually empty. This one wasn't. It groaned with wet Gore-Tex, and saturated sheepdog. I had inadvertently stumbled into Scilly's 'Twitch of the Day' – the curiously-named Temminck's Stint: a diminutive and hyperactive grey, brown-flecked, and buff-bellied wader that probed around frenetically at the water's edge. The cameras clicked. Back outside, quickly, the drizzle intensified. I dipped out of viewing a Lesser Yellowlegs from another hide – one silly name in a day is enough.

Next day, grey, out on the blasted heath of Wingletang, the exposed headland at the south-western end of St Agnes, the most westerly of the inhabited Scillies, existence became primeval, or downright elemental. Littered with granite men-hirs and giant boulders, whose lichen colonies seem to spell out some lost language, Wingletang is at best remorseless and unforgiving. In any gale it is evil.

Imagine, if you dare, the more brooding parts of high Dartmoor, where lingers a religion that existed at, or before, the dawn of man; upend that, granite altars, blood-stained quartz monoliths and all, and transport it by druidic power to the edge of the westernmost landmass known to the peoples of England; then add a wilder, more chaotic climate, and call it Wingletang. Do not even think of love here, even on a summer evening. This place is utterly amoral. This is Nature's kingdom, and it frightens the human mind. Beyond Wingletang lies 2,000 miles of emptiness, of endlessness. Keats beckoned:

> then on the shore
> Of the wide world I stand alone, and think
> Till love and fame to nothingness do sink.

John Keats - *When I Have Fears That I May Cease To Be*

The Meadow Pipits and departing Wheatears were forced to take shelter in hollows amongst boulder streams. Something sizeable and windblown hit me in the face, painfully. I've no idea what – it may not even have been of this earth, or of our time. It was meant to hurt, and had been aimed, by a valkyrie of autumn perhaps. At the most rational level, it was probably a frond of Bracken, plucked from the ground and hurled; for all around me the knee-high Bracken stems pirouetted helplessly: the gyrations of each stem had hollowed out the surrounding earth, penny-sized or more, which enabled them to whip around all the more, and all the more helplessly, hopelessly anchored, as a mighty metaphor for the human condition. Some broke loose, by mighty mishap.

Walking became nigh-impossible, for the wind blew footsteps out of kilter, so that boots landed other than where intended, and stumbled, madly and inanely. Clothing had become sailcloth, so that bodies could be driven on to jagged rocks and broken. All one could do was crouch behind a leaning menhir, and wait – and listen to the pitiless roar of another summer's destruction. What was going on in my mind was even more cataclysmic: I was becoming a grossly inferior version of poor Coleridge – an ailing metaphysician, haunted and hunted, inwardly.

Out at sea something more malevolent existed, beyond the curtain of horizontal rain and the casting of Gannets into the billows. The sea, when wild, seems scarcely of this world, not least its colours of nameless greys, white-tipped and foaming madly. Indeed, waves were erupting 50 feet high, over the jagged black rocks of Great Wingletang and westwards, far higher over Melledgan, the innermost of the treacherousness known as the Western Rocks. The spray blew on to Wingletang; such that it was impossible to breathe anything other than salt, and such that sea and air became one, unified by salt. The atmosphere, if such it was, reeked of winter. To the east, out towards the granite tors of Peninnis Point on St Mary's, is a belled buoy that tolls soullessly. Out here, on this day, it might drown.

Yet, amongst the Bent grasses within the lea of the monolith there were Violet leaves aplenty, telling of quieter, happier time; and amongst that vernal foliage were two or three purple flowers – next year's, already. And I knew then, whilst cramped down in the shelter of a granite slab that threatened to entomb me, alive or dead, that it was time – not so much to bid farewell to summer, and to ride out the pitiless waves of autumn, or deep-sleep the winter; but to look beyond all that drear and darkness, and commence the long dawning process of dreaming up the next spring – even as the Violets suggested; and on, to dream up a summer far greater than the one now passed, borne on a song-dream perfected by the love that comes from belonging in time and place, and nature. I was being called to ideate spring. So are we all.

There, helpless, out of the turbulence of a fell wind that was destroying summer, time became cathartic and I realised that my relationship with nature was not nearly sound enough. I decided then and there to ditch my job and re-wild myself – and dedicate the rest of my life, however long, to strengthening and deepening my relationship with Nature, and to hunt down the words that describe and share that wondrous bond, for Nature needs its spokespeople.

And I reached out in prayer and supplication beyond fore-seeable time, towards far-off spring and the long hot summer beyond, and held aloft next year's Violet flowers in my fingers. Yes, we hold in our hands the keys of nature, the words of life

– a flower handed to us as a pledge that we have walked through nature's Paradise, as if in a dream, and can walk there again:

And may the passion of the Blackbirds,
The perseverance of the Jackdaws,
The joy of the House Martins,
And the power of the Emperor's wings,
Be amongst you and within you,
Always…

Hugh Town, St Mary's, Isles of Scilly, 4th September.

Post Scriptum: The Soft-dying Day

The sky hangs o'er a broken dream
John Clare - *Decay: a Ballad*

A great poem speaks to us, not so much from yesterday, now, or tomorrow, but from almost beyond time. It matters not that archaic language may appear to anchor it in the past.

Perhaps great poems are written not so much by poets but by situations: most notably, by remarkable confluences of time, place, nature and mind; and that poets get used – wittingly hijacked – and become conduits through which time, place, circumstance and, most acutely, nature express themselves, through epiphanic experiences. This, in nature, can happen mightily, especially when our seasons are at their zenith. Indeed, nature surely uses its poets as its spokespeople. This is why the poetic approach to nature is so important.

That may well be what happened to John Keats on Sunday 19th September 1819, when he composed his ode *To Autumn*, and through it achieved the immortality he so ardently desired. This epiphany happened somewhere along the River Itchen, downstream of Winchester; past the cathedral, the college, and the ancient monastic hospice; through meadows, and past osiers that wind below St Catherine's Hill, with its plague pits and downland slopes; past the wavering of waterweed and Trout fins below the river's surface. He laid aside *The Fall of Hyperion*, an epic he had been struggling with for some time, for he was far more than a classical-ballad poet, and became his true self, a Prophet of Nature.

It was the culmination of a long hot summer that did it, finalised by Elysian September weather. The previous three summers had been dire, with failed harvests throughout Europe, most likely due to the impact on the climate of the eruption of the Tambora volcano in Indonesia early in 1815. By 1819 the dust had cleared from the stratosphere and a fine summer ensued, leading to a definitive September and a sound harvest. The consumptive Keats soared, both in spirit and physically.

Keats, though highly perceptive and responsive, was no ace naturalist, certainly not by today's standards. He was, after all, a cockney, albeit one who belonged out there in the natural world: for Nature claimed him, used him – and then immortalised him.

'How beautiful the season is now. How fine the air. A temperate sharpness about it... I never lik'd stubble fields as much as now...', he mused in his journal.

Then he broke through, or rather, something broke through, through him; and Keats found the words that best describe the fulfilment of summer which we call September, and through those words he became eternal, forever entwined with summer's glorification.

> Season of mists and mellow fruitfulness,
> Close bosom-friend of the maturing sun;
> Conspiring with him how to load and bless
> With fruit the vines that round the thatch-eves run;

To bend with apples the moss'd cottage-trees,
And fill all fruit with ripeness to the core;
To swell the gourd, and plump the hazel shells
With a sweet kernel; to set budding more,
And still more, later flowers for the bees,
Until they think warm days will never cease,
For Summer has o'er-brimm'd their clammy cells.

Who hath not seen thee oft amid thy store?
Sometimes whoever seeks abroad may find
Thee sitting careless on a granary floor,
Thy hair soft-lifted by the winnowing wind;
Or on a half-reap'd furrow sound asleep,
Drows'd with the fume of poppies, while thy hook
Spares the next swath and all its twined flowers:
And sometimes like a gleaner thou dost keep
Steady thy laden head across a brook;
Or by a cyder-press, with patient look,
Thou watchest the last oozings hours by hours.

Where are the songs of Spring? Ay, where are they?
Think not of them, thou hast thy music too,—
While barred clouds bloom the soft-dying day,
And touch the stubble plains with rosy hue;
Then in a wailful choir the small gnats mourn
Among the river sallows, borne aloft
Or sinking as the light wind lives or dies;
And full-grown lambs loud bleat from hilly bourn;
Hedge-crickets sing; and now with treble soft
The red-breast whistles from a garden-croft;
And gathering swallows twitter in the skies

John Keats - *To Autumn*

September 21st, St Matthew's Day.

www.ingramcontent.com/pod-product-compliance
Lightning Source LLC
Chambersburg PA
CBHW022332280326
41934CB00006B/602